T0269248

LONDON MATHEMATICAL SOCIETY LECTURE NOTE SERIES

Managing Editor: Professor J.W.S. Cassels, Department of Pure Mathematics and Mathematical Statistics, University of Cambridge, 16 Mill Lane, Cambridge CB2 1SB, England

The titles below are available from booksellers, or, in case of difficulty, from Cambridge University Press.

London Mathematical Society Lecture Note Series. 194

Independent Random Variables and Rearrangement Invariant Spaces

Michael Sh. Braverman
University of Khabarovsk

CAMBRIDGE
UNIVERSITY PRESS

Published by the Press Syndicate of the University of Cambridge
The Pitt Building, Trumpington Street, Cambridge CB2 1RP
40 West 20th Street, New York, NY 10011-4211, USA
10 Stamford Road, Oakleigh, Melbourne 3166, Australia

First published 1994

Library of Congress cataloguing in publication data available

British Library cataloguing in publication data available

ISBN   0 521 45515 4 paperback

Transferred to digital printing 2004

CONTENTS

PREFACE

Rearrangement invariant (r.i.) spaces were first considered in connection with questions arising in the theory of interpolation of linear operators. Now they are the object of intensive study. The main property of r.i. spaces is that the norm of any element X depends only on the distribution of X. This is why probabilistic methods have been used successfully.

The purpose of this monograph is to study sequences of independent random variables (r.v.) as sequences of elements of r.i. spaces. Similar problems have been discussed by many authors (see [15], [17], [23], [24], [33] et al.). The work consists of four chapters. The first chapter contains some well known notation and results from probability theory and the theory of r.i. spaces which are used in the following chapters. Besides, we introduce a new term, which we will call the Kruglov property.

Let F be a probability distribution on $\mathbf{R}$ and $\Pi(F)$ be the corresponding compound Poisson distribution, i.e. the distribution with the characteristic function

$$g(t) = \exp\left(\int_{-\infty}^{\infty}(e^{itx}-1)dF(x)\right).$$

Kruglov [30] described the class of positive functions $\Phi(x)$ on $\mathbf{R}$ for which the conditions

$$\int_{-\infty}^{\infty}\Phi(x)dF(x) < \infty, \quad \int_{-\infty}^{\infty}\Phi(x)d(\Pi(F))(x) < \infty$$

are equivalent. We say that a r.i. space $\mathbf{E}$ has the Kruglov property ($\mathbf{E} \in \mathbf{K}$) if for any r.v.s X and Y with the distributions F and $\Pi(F)$ respectively the conditions $X \in \mathbf{E}$ and $Y \in \mathbf{E}$ are equivalent.

For instance, $L_p \in \mathbf{K}$ if $p < \infty$. The Kruglov property plays an important role in our considerations because it permits us to compare sequences of independent and disjoint r.v.s.

In Chapter 2 we prove several results about analogs of the Rosenthal and the von Bahr and Esseen inequalities (see [2] and [50]). One of them is the following: if $\mathbf{E} \in \mathbf{K}$ and some analog of Rosenthal's inequality holds, then $\mathbf{E} = L_p$ for some $p \in (2, \infty)$.

Another result states that under the assumption $\mathbf{E} \in \mathbf{K}$ an analog of the von Bahr and Esseen estimate is fulfilled in $\mathbf{E}$ if and only if a similar inequality is true for mutually disjoint r.v.s, i.e. $\mathbf{E}$ satisfies the upper p-estimate. We note that similar results have been proved in [14] and [24].

In Chapter 3 we study sequences of independent identically distributed r.v.s. The results on the conditions under which such a sequence generates the subspace in $\mathbf{E}$ isomorphic to the space l_p are obtained.

In Chapter 4 the question of the complementability in **E** of the subspace generated by a sequence of independent r.v.s is considered. Similar problems have been studied by many authors (see [**17**], [**33**], [**49**] et al.).

Each chapter has its own enumeration. With references to a statement or formula of another chapter the number of the latter is assigned, but it is not added if the reference is to the same chapter.

CHAPTER I

PRELIMINARIES

1. Rearrangement invariant spaces

Let $(\Omega, \mathcal{F}, P)$ be a probability space, i.e. a measure space such that $P(\Omega) = 1$. A measurable function X on Ω is called a *random variable* (*r.v.*). We write $X \overset{d}{=} Y$ if these r.v.s are identically distributed. This means that $P\{X < x\} = P\{Y < x\}$ for every $x \in \mathbf{R}$. The *mean value* of a r.v. X is defined by the formula

$$EX = \int_\Omega X(\omega) dP(\omega).$$

The *distribution* of a r.v. X is the measure F on $\mathbf{R}$ defined by the formula $F(h) = P\{X \in h\}$ for every measurable set $h \subset \mathbf{R}$. We write $X \in \mathcal{L}(F)$ if X has the distribution F. The function

$$f(t) = \int_{-\infty}^{\infty} \exp(itx) dF(x)$$

is called the *characteristic function* of the r.v. X.

Definition 1. *A Banach space* $\mathbf{E}$ *of random variables defined on* $(\Omega, \mathcal{F}, P)$ *is said to be rearrangement invariant* (*r.i.*) *if the following conditions hold:*
  (i) *if* $|X| \leq |Y|$ *and* $Y \in \mathbf{E}$, *then* $X \in \mathbf{E}$ *and* $\|X\|_{\mathbf{E}} \leq \|Y\|_{\mathbf{E}}$;
  (ii) *if* $X \overset{d}{=} Y$ *and* $Y \in \mathbf{E}$, *then* $X \in \mathbf{E}$ *and* $\|X\|_{\mathbf{E}} = \|Y\|_{\mathbf{E}}$.

The main information on r.i. spaces is contained in the book [29]. The spaces $L_p(\Omega)\,(1 \leq p \leq \infty)$ are rearrangement invariant. The Lorentz spaces $L_{p,q}(\Omega)$ form a wider class of r.i. spaces which consist of all r.v.s X such that

$$\|X\|_{p,q}^* = \left(\int_0^\infty (P\{|X| \geq x\})^{q/p} dx^q \right)^{1/q} < \infty \tag{1}$$

if $1 \leq p, q < \infty$, and,

$$\|X\|_{p,\infty}^* = \sup \left\{ x(P\{|X| \geq x\})^{1/p} : x > 0 \right\} < \infty. \tag{2}$$

The functionals (1) and (2) are not norms in general, but they are equivalent to some norms (see [52]). We have $L_{p,p} = L_p$.

Orlicz spaces form another class of r.i. spaces. 1Let $N(x)$ be a convex even function on $\mathbf{R}$, $N(0) = 0$. The Orlicz space $L_N(\Omega)$ consists of all r.v.s X such that $EN(\lambda^{-1}X) < \infty$ for some $\lambda > 0$. The norm is defined by the formula (see [28])

$$\|X\|_{L_N} = \inf\left\{\lambda > 0 : EN(\lambda^{-1}X) < 1\right\}. \qquad (3)$$

In the sequel the probability space $(\Omega, \mathcal{F}, P)$ is assumed to be non-atomic. The *indicator* of a set h is denoted by I_h.

The following statements are proved in [29].

Proposition 1. *For every r.i. space* $\mathbf{E}$

$$L_\infty(\Omega) \subset \mathbf{E} \subset L_1(\Omega)$$

and for each r.v. $X \in L_\infty(\Omega)$

$$\|X\|_{L_1(\Omega)} \leq \frac{\|X\|_{\mathbf{E}}}{\|I_\Omega\|_{\mathbf{E}}} \leq \|X\|_{L_\infty(\Omega)}.$$

Proposition 2. *Let* $Y \in \mathbf{E}$ *and* $P\{|X| \geq x\} \leq CP\{|Y| \geq x\}$ *for all* $x > 0$, *where* C *is a constant. Then* $X \in \mathbf{E}$ *and* $\|X\|_{\mathbf{E}} \leq \max\{1, C\}\|Y\|_{\mathbf{E}}$.

The *dual* or *associated* space $\mathbf{E}'$ of the r.i. space $\mathbf{E}$ is the set of all r.v.s Y such that

$$\|Y\|_{\mathbf{E}'} \overset{def}{=} \sup\left\{EXY : X \in \mathbf{E}, \|X\|_{\mathbf{E}} \leq 1\right\} < \infty.$$

It is well known (see [29]) that $\mathbf{E}'$ is a r.i. space. Write $\mathbf{E}'' = (\mathbf{E}')'$. We have $\mathbf{E} \subset \mathbf{E}''$ and $\|X\|_{\mathbf{E}''} \leq \|X\|_{\mathbf{E}}$ for every $X \in \mathbf{E}$. If $X \in L_\infty(\Omega)$, then $\|X\|_{\mathbf{E}''} = \|X\|_{\mathbf{E}}$. A r.i. space $\mathbf{E}$ is said to be *maximal*, if $\mathbf{E}'' = \mathbf{E}$.

The *decreasing rearrangement* of a r.v. X is the function on $(0,1)$ defined by the formula

$$X^*(t) = \inf\left\{s > 0 : P\{|X| \geq s\} < t\right\}.$$

We write $X \prec Y$ for $X, Y \in L_1(\Omega)$ if for each $0 \leq t \leq 1$

$$\int_0^t X^*(s)\,ds \leq \int_0^t Y^*(s)\,ds.$$

The following assertions are proved in [29].

Proposition 3. *Let a r.i. space* $\mathbf{E}$ *be maximal or separable,* $Y \in \mathbf{E}$ *and* $X \prec Y$. *Then* $X \in \mathbf{E}$ *and* $\|X\|_{\mathbf{E}} \leq \|Y\|_{\mathbf{E}}$.

Proposition 4 (Calderon–Mityagin's Theorem). *Suppose that a r.i. space* $\mathbf{E}$ *has the following property: if* $Y \in \mathbf{E}$ *and* $X \prec Y$ *then* $X \in \mathbf{E}$ *and*

$\|X\|_{\mathbf{E}} \leq \|Y\|_{\mathbf{E}}$. *Let a linear operator* T *be bounded in* $L_1(\Omega)$ *and* $L_\infty(\Omega)$. *Then* T *is bounded in* **E** *and*

$$\|T\|_{\mathbf{E}\to\mathbf{E}} \leq \max\left\{\|T\|_{L_1(\Omega)\to L_1(\Omega)}, \ \|T\|_{L_\infty(\Omega)\to L_\infty(\Omega)}\right\}.$$

Let X_n and X be r.v.s with the distributions F_n and F respectively. The sequence X_n is said to be *weakly convergent to* X if for every continuous bounded function $g(x)$ on **R**

$$\lim_{n\to\infty} \int_{-\infty}^{\infty} g(x)dF_n(x) = \int_{-\infty}^{\infty} g(x)dF(x).$$

Weak convergence is equivalent to each of the following conditions:
(i) $F_n(x) \to F(x)$ for all $x \in \mathbf{R}$ on which F is continious ;
(ii) $f_n(t) \to f(t)$ for each $t \in \mathbf{R}$, where $f_n(t)$ and $f(t)$ are the corresponding characteristic functions (see [51]).

Proposition 5. *Let* **E** *be a maximal r.i. space and* $X_n \in \mathbf{E}$. *Suppose* X_n *is weakly convergent to* X *and*

$$\sup_n \|X_n\|_{\mathbf{E}} = C < \infty.$$

Then $X \in \mathbf{E}$ *and* $\|X\|_{\mathbf{E}} \leq C$.

Proof: Let $Y \in \mathbf{E}'$ and $\|Y\|_{\mathbf{E}'} \leq 1$. According to Proposition 1, $E|Y| < \infty$. From here $Y^* \in L_1(0,1)$ and the function

$$\Psi_Y(t) = \int_0^t Y^*(s)ds$$

is well defined on $(0,1)$. It is clear that

$$\int_0^1 X^*(t)Y^*(t)dt = \int_0^1 X^*(t)d\Psi_Y(t).$$

We use the formulae (see [29], Ch. 2)

$$\int_0^1 X^*(t)d\Psi_Y(t) = \int_0^\infty \Psi_Y(P\{|X| \geq x\})dx$$

and

$$\|X\|_{\mathbf{E}''} = \sup\left\{\int_0^1 X^*(t)Y^*(t)dt : \|Y\|_{\mathbf{E}'} \leq 1\right\}.$$

Fatou's lemma and these equalities imply that

$$\int_0^\infty \Psi_Y(P\{|X| \geq x\})dx \leq \sup_n \int_0^\infty \Psi_Y(P\{|X_n| \geq x\})dx$$

$$= \sup_n \int_0^1 X_n^*(t)Y^*(t)dt \leq \sup_n \|X_n\|_{\mathbf{E}} = C.$$

So,

$$\sup\left\{ \int_0^1 X^*(t)Y^*(t)dt \; : \; \|Y\|_{\mathbf{E}'} \leq 1 \right\} \leq C < \infty.$$

Hence $X \in \mathbf{E}'' = \mathbf{E}$ and $\|X\|_{\mathbf{E}} \leq C.$ $\square$

If $\mathbf{E}$ is a r.i. space, then the norm $\|I_h\|_{\mathbf{E}}$ depends on $P(h)$ only. Therefore the function

$$\phi_{\mathbf{E}}(t) = \|I_h\|_{\mathbf{E}} \quad (P(h) = t)$$

is well defined on $[0, 1]$. It is called *the fundamental function* of $\mathbf{E}$. The next statement follows from the results of [**29**], Ch. 2.

Proposition 6. *If*

$$b_{\mathbf{E}}(X) \overset{def}{=} \int_0^\infty \phi_{\mathbf{E}}(P\{|X| \geq x\})dx < \infty,$$

then $X \in \mathbf{E}$ and $\|X\|_{\mathbf{E}} \leq a(\mathbf{E})b_{\mathbf{E}}(X)$, where $a(\mathbf{E})$ depends on $\mathbf{E}$ only.

2. The function of dilatation and Boyd indices

For the r.v. X and $t > 0$ let $B_t(X)$ denote the set of all r.v.s Y such that $P\{x \leq Y < y\} \leq tP\{x \leq X < y\}$ for every $x < y, xy > 0$. If $Y \in B_t(X)$, then

$$P\{|Y| \geq x\} \leq tP\{|X| \geq x\}$$

for all positive x. Proposition 2 implies that if $X \in \mathbf{E}$, then $B_t(X) \subset \mathbf{E}$ for every $t > 0$. Put

$$\gamma_{\mathbf{E}}(t) = \sup\left\{ \frac{\|X\|_{\mathbf{E}}}{\|Y\|_{\mathbf{E}}} \; : \; Y \in B_t(X), \; X \in \mathbf{E}, \; X \neq 0 \right\}.$$

We call $\gamma_{\mathbf{E}}(t)$ the *function of dilatation* of the r.i. space $\mathbf{E}$. It is obvious that $\gamma_{\mathbf{E}}(t)$ is non-decreasing.

Suppose that the probability space is $[0, 1]$ with Lebesgue measure and for a r.v. $X(s)$ on $[0, 1]$ putp

$$D_t X(s) = \begin{cases} X(s/t) & \text{if } s < t, \\ 0 & \text{if } t < s \leq 1. \end{cases}$$

The operator D_t is bounded in every r.i. space $\mathbf{E}$ (see [**29**]). It is easy to verify that $\gamma_{\mathbf{E}}(t) = \|D_t\|_{\mathbf{E} \to \mathbf{E}}.$

Proposition 7. *The following estimates are true:*

$$\min\{1,t\} \le \gamma_{\mathbf{E}}(t) \le \max\{1,t\}.$$

These inequalities follow directly from the definition and the obvious relations $\gamma_{\mathbf{E}}(t) \le 1 = \gamma_{\mathbf{E}}(1)$ if $t \le 1$ and $\gamma_{\mathbf{E}}(t) \ge 1$ if $t > 1$.

The *Boyd indices* are defined by the formulae

$$\alpha(\mathbf{E}) = \lim_{t\to 0} \frac{\log(\gamma_{\mathbf{E}}(t))}{\log(t)}, \quad \beta(\mathbf{E}) = \lim_{t\to\infty} \frac{\log(\gamma_{\mathbf{E}}(t))}{\log(t)}.$$

These limits exist and $0 \le \alpha(\mathbf{E}) \le \beta(E) \le 1$ (see [29], Ch.2). We have $\alpha(L_{p,q}) = \beta(L_{p,q}) = 1/p$.

Proposition 8. *For each $\epsilon > 0$ there exist positive constants a and b such that $\gamma_{\mathbf{E}}(t) \le t^{\alpha(\mathbf{E})-\epsilon}$ if $t < a$ and $\gamma_{\mathbf{E}}(t) \le t^{\beta(\mathbf{E})+\epsilon}$ if $t > b$.*

Proof: See [29], Ch. 2. □

The following assertion is well known (see [23]).

Proposition 9. *If $\alpha(\mathbf{E}) > 0$, then $\mathbf{E} \supset L_p(\Omega)$ for all $p > 1/\alpha(\mathbf{E})$. If $\beta(\mathbf{E}) < 1$, then $\mathbf{E} \subset L_q(\Omega)$ for each $q < 1/\beta(\mathbf{E})$.*

3. Independent random variables

Definition 2. *Let H be a set of indices. The r.v.s $\{X_h\}$, $h \in H$, are said to be independent, if for every finite set $h_1,\ldots,h_m$ and each $a_k < b_k$ ($1 \le k \le m$)*

$$P\{a_k \le X_{h_k} \le b_k, 1 \le k \le m\} = \prod_{k=1}^{m} P\{a_k \le X_{h_k} \le b_k\}.$$

Proposition 10. *Let r.v.s X and Y be independent, $\mathbf{E}$ be a r.i. space and $X + Y \in \mathbf{E}$. Then $X \in \mathbf{E}$ and $Y \in \mathbf{E}$.*

Proof: Let α be a *median* of X. It means $P\{X \le \alpha\} \ge 1/2$ and $P\{X \ge \alpha\} \ge 1/2$ (see [35]). Since X and Y are independent, then for every $x > 0$

$$P\{|X + Y| \ge x\}$$

$$\ge P\{X \ge x - \alpha, Y \ge \alpha\} + P\{X \le -x - \alpha, Y \le \alpha\}$$

$$\ge \frac{1}{2}P\{X \ge x - \alpha\} + \frac{1}{2}P\{X \le -x - \alpha\}$$

$$\ge \frac{1}{2}P\{|X| \ge x + |\alpha|\}.$$

Using Proposition 2, we get $X \in \mathbf{E}$. □

Proposition 11. *For each r.i. space* $\mathbf{E}$ *there exists a constant* $C(\mathbf{E}) > 0$ *such that for any independent r.v.s* $X, Y \in \mathbf{E}$, $EY = 0$

$$\|X + Y\|_{\mathbf{E}} \geq C(\mathbf{E})\|X\|_{\mathbf{E}}.$$

Proof: Suppose that this assertion is not true for some r.i. space $\mathbf{E}$. Then there exist r.v.s $X_n, Y_n \in \mathbf{E}$ with the following properties:
1) the r.v.s $\{X_k\}_{k=1}^{\infty} \bigcup \{Y_k\}_{k=1}^{\infty}$ are independent;
2) $EY_k = 0$;
3) $\|X_k\|_{\mathbf{E}} = 1$;
4) $\|X_k + Y_k\|_{\mathbf{E}} \leq 2^{-k}$.

According to 4), the series $\sum_{k=1}^{\infty}(X_k + Y_k)$ is absolutely convergent in $\mathbf{E}$ and, therefore, in $L_1(\Omega)$. Let β be the σ-algebra generated by the sequence $\{X_k\}_{k=1}^{\infty}$. Since the operator E^{β} of conditional expectation is bounded in $L_1(\Omega)$, the conditions 1) and 2) imply for each $m < n$ (see [35])

$$E^{\beta}\left(\sum_{k=m}^{n}(X_k + Y_k)\right) = \sum_{k=m}^{n} X_k + \sum_{k=m}^{n} EY_k = \sum_{k=m}^{n} X_k.$$

Hence the series $\sum_{k=1}^{\infty} X_k$ and $\sum_{k=1}^{\infty} Y_k = \sum_{k=1}^{\infty}(X_k + Y_k) - \sum_{k=1}^{\infty} X_k$ are convergent in $L_1(\Omega)$. According to 1) and 2), the second series is convergent almost surely (see [35]). Hence $Y_k \to 0$ almost surely.

Let $h_k(\epsilon) = \{|Y_k| < \epsilon\}$, where $0 < \epsilon < 1$. We have $P(h_k(\epsilon)) \geq 1 - \epsilon$ for sufficiently large k and

$$|X_k + Y_k|I_{h_k(\epsilon)} \geq \||X_k| - \epsilon|I_{h_k(\epsilon)}.$$

By virtue of independence, for each $x > 0$

$$P\left\{\||X_k| - \epsilon|I_{h_k(\epsilon)} \geq x\right\} = P(h_k(\epsilon))P\left\{\||X_k| - \epsilon| \geq x\right\}.$$

From here and Proposition 2

$$\|(|X_k| - \epsilon)I_{h_k(\epsilon)}\|_{\mathbf{E}} \geq P(h_k(\epsilon))\||X_k| - \epsilon\|_{\mathbf{E}}.$$

Hence for sufficiently large k

$$\begin{aligned}
\|X_k + Y_k\|_{\mathbf{E}} &\geq \|(X_k + Y_k)I_{h_k(\epsilon)}\|_{\mathbf{E}} \geq \|(|X_k| - \epsilon)I_{h_k(\epsilon)}\|_{\mathbf{E}} \\
&\geq P(h_k(\epsilon))\||X_k| - \epsilon\|_{\mathbf{E}} \geq (1 - \epsilon)(\|X_k\|_{\mathbf{E}} - \epsilon) \\
&= (1 - \epsilon)^2.
\end{aligned}$$

This contradicts 4). $\square$

Remark. If **E** is separable or maximal, then $C(\mathbf{E}) = 1$. Indeed, let β be the σ-algebra generated by X. Proposition 4 implies that the operator E^β is bounded in **E** and has norm equal to 1. Hence

$$\|X + Y\|_\mathbf{E} \geq \|E^\beta(X + Y)\|_\mathbf{E} = \|X + EY\|_\mathbf{E} = \|X\|_\mathbf{E}.$$

For the spaces $L_p(\Omega)$ $(1 \leq p \leq \infty)$ this statement is well known (see [35]).

4. Probability inequalities

Here we present some well known estimates which will be used in the following chapters.

1. Paley and Zygmund's inequality [26]. *Let* $\{X_k\}_{k=1}^\infty$ *be independent r.v.s such that* $EX_k^4 < \infty$ *and* $EX_k = 0$. *Suppose*

$$D = \sup \left\{ \frac{EX_k^4}{(EX_k^2)^2} < \infty \right\}.$$

Then for every $a_k \in \mathbf{R}$ *and* $n \in \mathbf{N}$

$$P\left\{ \left| \sum_{k=1}^n a_k X_k \right| \geq \lambda \left(\sum_{k=1}^n a_k^2 \right)^{1/2} \right\} \geq \eta,$$

where $\eta = (1 - \lambda^2) \min\{1/3, 1/D\}$.

2. Bernstein's inequality [43]. *Let* $\{X_k\}_{k=1}^\infty$ *be independent r.v.s,* $|X_k| \leq C$ *and* $EX_k = 0$. *Then*

$$P\left\{ \left| \sum_{k=1}^n a_k X_k \right| \geq 2Cx \left(\sum_{k=1}^n a_k^2 \right)^{1/2} \right\} \leq \exp(-x^2)$$

for all $x > 0$, $a_k \in \mathbf{R}$ *and* $n \in \mathbf{N}$.

3. Prokhorov's 'arcsinh' inequality [46]. *Let* $\{X_k\}_{k=1}^\infty$ *be independent r.v.s,* $|X_k| \leq C$ *and* $EX_k = 0$. *Let* $\sigma^2 = \sum_{k=1}^n EX_k^2$. *Then for every* $x > 0$

$$P\left\{ \left| \sum_{k=1}^n X_k \right| \geq x \right\} \leq 2\exp\left(-\frac{x}{2C} \operatorname{arcsinh} \frac{Cx}{\sigma^2} \right).$$

A r.v. X is said to be *symmetric* if $P\{X \geq x\} = P\{X \leq -x\}$ for all $x > 0$. If F is a probability distribution on **R**, then $\Pi(F)$ denotes the corresponding Poisson distribution, i.e. the distribution with the characteristic function

$$g(t) = \exp\left(\int_{-\infty}^\infty (e^{itx} - 1) dF(x) \right). \tag{4}$$

Let us recall we write $X \in \mathcal{L}(F)$ if F is the distribution of the r.v. X.

4. Prokhorov's inequality [45]. *Let $\{X_k\}_{k=1}^n$ be independent symmetric r.v.s with distributions F_k and let the r.v.s $Y_k \in \mathcal{L}(\Pi(F_k))$ be independent. Then for every $x > 0$*

$$P\left\{\left|\sum_{k=1}^n X_k\right| \geq x\right\} \leq 8P\left\{\left|\sum_{k=1}^n Y_k\right| \geq \frac{x}{2}\right\}.$$

5. Kwapien and Richlik's inequality [53]. *Let $\{X_k\}_{k=1}^\infty$ and $\{Y_k\}_{k=1}^\infty$ be sequences of independent symmetric r.v.s such that $P\{|X_k| \geq x\} \leq AP\{B|Y_k| \geq x\}$ for all $x > 0$ and $k \in \mathbf{N}$. Then for every $a_k \in \mathbf{R}$, $x > 0$ and $n \in \mathbf{N}$*

$$P\left\{\left|\sum_{k=1}^n a_k X_k\right| \geq x\right\} \leq 2AP\left\{AB\left|\sum_{k=1}^n a_k Y_k\right| \geq x\right\}.$$

5. Disjoint random variables

We say the r.v.s X and Y are *disjoint*, if $XY = 0$.

Definition 3. *[33]. A r.i. space $\mathbf{E}$ is said satisfy the upper, respectively the lower, r-estimate if there exists a positive constant A such that for each collection of mutually disjoint r.v.s $\{X_k\}_{k=1}^n \subset \mathbf{E}$*

$$\left\|\sum_{k=1}^n X_k\right\|_{\mathbf{E}} \leq A\left(\sum_{k=1}^n \|X_k\|_{\mathbf{E}}^r\right)^{1/r},$$

respectively

$$\left\|\sum_{k=1}^n X_k\right\|_{\mathbf{E}} \geq A\left(\sum_{k=1}^n \|X_k\|_{\mathbf{E}}^r\right)^{1/r}.$$

The following assertions are known. For completeness the proofs are given.

Proposition 12. *Let $1 \leq p < \infty$, $1 \leq q \leq \infty$ and $r = \min\{p,q\}$. The Lorentz space $L_{p,q}(\Omega)$ satisfies the upper r-estimate and does not satisfy the upper s-estimate for $s > r$.*

Proof: Let $\{X_k\}_{k=1}^n \subset \mathbf{E}$ be mutually disjoint r.v.s. Then for every $x > 0$

$$P\left\{\left|\sum_{k=1}^n X_k\right| \geq x\right\} = \sum_{k=1}^n P\{|X_k| \geq x\}.$$

Suppose $q = \infty$. According to (2)

$$\left(\left\|\sum_{k=1}^{n} X_k\right\|_{p,\infty}^{*}\right)^p = \sup\left\{x^p P\left\{\left|\sum_{k=1}^{n} X_k\right| \geq x\right\} : x > 0\right\}$$

$$= \sup\left\{x^p \sum_{k=1}^{n} P\{|X_k| \geq x\} : x > 0\right\}$$

$$\leq \sum_{k=1}^{n} \left(\|X_k\|_{p,\infty}^{*}\right)^p.$$

Let $r = p \leq q < \infty$. From the formula (1)

$$\left(\left\|\sum_{k=1}^{n} X_k\right\|_{p,q}^{*}\right)^p = \left(\int_{0}^{\infty} \left(\sum_{k=1}^{n} P\{|X_k| \geq x\}\right)^{q/p} dx^q\right)^{p/q}.$$

Applying Minkowski's inequality with the exponent $\alpha = q/p$, we get the needed estimate.

Let's turn to the case $r = q < p$. We have

$$\left(\left\|\sum_{k=1}^{n} X_k\right\|_{p,q}^{*}\right)^q = \int_{0}^{\infty} \left(\sum_{k=1}^{n} P\{|X_k| \geq x\}\right)^{q/p} dx^q.$$

Since $|a + b|^t \leq |a|^t + |b|^t$ for $0 < t < 1$, then

$$\left(\sum_{k=1}^{n} P\{|X_k| \geq x\}\right)^{q/p} \leq \sum_{k=1}^{n} (P\{|X_k| \geq x\})^{q/p}.$$

From here the desired estimate follows.

Let $s > r$. If a r.i. space $\mathbf{E}$ satisfies the upper s-estimate, then $\beta(\mathbf{E}) < 1/s$ (see [33]). Since $\beta(L_{p,q}) = 1/p$, then, if $r = p$, the space $L_{p,q}(\Omega)$ does not satisfy the upper s-estimate.

Suppose $r = q$. Then, as shown in [40], there exist mutually disjoint normed r.v.s $\{X_k\}_{k=1}^{\infty} \subset L_{p,q}(\Omega)$ such that for all $n \in \mathbf{N}$

$$\left\|\sum_{k=1}^{n} X_k\right\|_{p,q}^{*} \geq n,$$

where $C > 0$ is a constant. So, $L_{p,q}(\Omega)$ does not satisfy the upper s-estimate for $s > r = q$ and the proposition is proved. $\square$

Now we consider Orlicz spaces. Let $N_i(x)$ ($i = 1, 2$) be even convex functions on $\mathbf{R}$. We say that these functions are *equivalent* if there exist positive constants A, B and C such that $N_1(Ax) \leq N_2(x) \leq N_1(Bx)$ for $x > C$. It is shown in [28] that $L_{N_1}(\Omega) = L_{N_2}(\Omega)$ if and only if the functions $N_1(x)$ and $N_2(x)$ are equivalent.

Proposition 13. *Let a convex even function $N(x)$, $N(0) = 0$, be equivalent to $U(x) = |x|^r \Psi(x)$, where $r > 1$ and $\Psi(x)$ is convex and increasing on $(0, \infty)$. Then the space $L_N(\Omega)$ satisfies the upper r-estimate.*

Proof.: Without loss of generality $N(x) = |x|^r \Psi(x)$. Suppose $\{X_k\}_{k=1}^n \subset L_N(\Omega)$ are mutually disjoint random variables and $EN(\lambda_k^{-1} X_k) \leq 1$ for some real λ_k ($1 \leq k \leq n$). Put $t_k = \lambda_k (\sum_{j=1}^n \lambda_j^r)^{-1/r}$. Then

$$EN\left(\left(\sum_{j=1}^n \lambda_j^r\right)^{-1/r} \sum_{k=1}^n X_k\right) = \sum_{k=1}^n EN(t_k \lambda_k^{-1} X_k).$$

Since the function $\Psi(x)$ is even and increasing on $(0, \infty)$ we have $\Psi(tx) \leq \Psi(x)$ for $0 < t < 1$ and every $x \in \mathbf{R}$. Therefore $EN(tX) = E(|tX|^r \Psi(tX)) \leq |t|^r E(|X|^r \Psi(X)) = |t|^r EN(X)$. From here

$$EN\left(\left(\sum_{j=1}^n \lambda_j^r\right)^{-1/r} \sum_{k=1}^n X_k\right) \leq \sum_{k=1}^n t_k^r EN(\lambda_k^{-1} X_k) \leq \sum_{k=1}^n t_k^r = 1.$$

Using (3), we get

$$\left\| \sum_{k=1}^n X_k \right\|_{L_N} \leq \left(\sum_{k=1}^n t_k^r \right)^{1/r}.$$

Applying (3) once more, we obtain the desired bound. $\square$

6. The Kruglov property

In this section we investigate relations between independent r.v.s and mutually disjoint ones. A similar problem has been studied by Carothers and Dilworth [14], [15] and Johnson and Schechtman [24]. Our approach is different from theirs. It is based on the so-called Kruglov property of a r.i. space.

Kruglov proved the following statement [30].

Kruglov's Theorem. *Let $\Phi(x)$ be a non-negative and continuous function on $\mathbf{R}$ satisfying one of the following conditions:*

$$\Phi(x + y) \leq B\Phi(x)\Phi(y), \quad \Phi(x + y) \leq B(\Phi(x) + \Phi(y)),$$

where B is a constant. Suppose $X \in \mathcal{L}(F)$ and $Y \in \mathcal{L}(\Pi(F))$. Then the conditions $E\Phi(X) < \infty$ and $E\Phi(Y) < \infty$ are equivalent.

The question arises about the conditions under which the result of this type is true for r.i. spaces.

Definition 4. *We say that a r.i. space* $\mathbf{E}$ *has the Kruglov property* ($\mathbf{E} \in \mathbf{K}$) *if for* $X \in \mathcal{L}(F)$ *and* $Y \in \mathcal{L}(\Pi(F))$ *the conditions* $X \in \mathbf{E}$ *and* $Y \in \mathbf{E}$ *are equivalent.*

From Kruglov's Theorem, $L_p(\Omega) \in \mathbf{K}$ for $1 \leq p < \infty$.

We prove some results about the Kruglov property which will be used in the following chapters.

Lemma 1. *Let* $X \in \mathcal{L}(F)$ *and* $Y \in \mathcal{L}(\Pi(F))$. *Let* $X \in \mathbf{E}$ *and* $\mathbf{E} \in \mathbf{K}$. *Then*

$$e^{-1}\|X\|_{\mathbf{E}} \leq \|Y\|_{\mathbf{E}} \leq C_1(\mathbf{E})\|X\|_{\mathbf{E}}, \tag{5}$$

where $C_1(\mathbf{E})$ *is a constant.*

Proof: The following well known construction is used. Let $h_n \subset \Omega$ ($n = 0, 1, \dots$) be mutually disjoint sets and $P(h_n) = 1/(n!e)$. Consider r.v.s $X_{k,n}$ ($1 \leq k \leq n$, $n = 1, 2, \dots$) such that
1) the r.v.s $\{X_{k,n}\}_{k=1}^n \bigcup \{I_{h_n}\}$ are independent for every n ;
2) $X_{k,n} \stackrel{d}{=} X$ ($1 \leq k \leq n$, $n = 1, 2, \dots$). Put

$$Y = \sum_{n=1}^{\infty} \left(\sum_{k=1}^{n} X_{k,n} \right) I_{h_n}. \tag{6}$$

It is easily verified that the characteristic function of Y is given by the formula (4). Hence $Y \in \mathcal{L}(\Pi(F))$.

We have from (6) $P\{|Y| \geq x\} \geq e^{-1}P\{|X| \geq x\}$ for all $x > 0$. So, the left-hand side of (5) follows.

Let's turn to the right-hand side of (5). Suppose it is not true. Then there exist r.v.s $Z_j \in \mathbf{E}$ ($j = 1, 2, \dots$) with distributions G_j such that $\|Y_j\|_{\mathbf{E}} \geq 2^{2j}\|Z_j\|_{\mathbf{E}}$, where $Y_j \in \mathcal{L}(\Pi(G_j))$.

Let G and H be the distributions of the r.v.s X and $|X|$ respectively. Replacing in (6) $X_{k,n}$ by $|X_{k,n}|$, we get the r.v. U with the distribution $\Pi(H)$. It is obvious that $|Y| \leq U$. Since X and $|X|$ have the same norm in $\mathbf{E}$, then the previous inequality also holds for the r.v.s $|Z_j|$.

Let G_a be the distribution of aX, where $a \in \mathbf{R}$. It follows from (6), that if $Y \in \mathcal{L}(\Pi(G))$ and $Y_a \in \mathcal{L}(\Pi(G_a))$, then $aY \stackrel{d}{=} Y_a$. Multiplying $|Z_j|$ by suitable constants, we get the r.v.s X_j with the distributions F_j such that

$$X_j \geq 0, \quad \|X_j\|_{\mathbf{E}} \leq 2^{-j}, \quad \|Y_j\|_{\mathbf{E}} \geq 2^j, \tag{7}$$

where $Y_j \in \mathcal{L}(\Pi(G_j))$. We may suppose the r.v.s $\{X_j\}_{j=1}^{\infty}$ to be independent. Put

$$X = \sum_{j=1}^{\infty} X_j.$$

According to (7), this series is absolutely convergent in $\mathbf{E}$ and, therefore, $X \in \mathbf{E}$. Let F be the distribution of X. We show that $\mathbf{E}$ does not contain a r.v. Y with the distribution $\Pi(F)$, which contradicts the Kruglov property.

Proposition 1 yields that the considered series is absolutely convergent in $L_1(\Omega)$. Therefore the series $\sum_{k=1}^{\infty}(X_j - EX_j)$ is convergent in $L_1(\Omega)$ also. Since the r.v.s $\{X_j\}_{j=1}^{\infty}$ are independent, the last series is convergent almost surely (see [35]). Hence the series $\sum_{k=1}^{\infty} X_j$ is also convergent almost surely.

Let $\{h_n\}_{n=0}^{\infty}$ be the same sets as above. We may choose r.v.s $X_{j,k,n}$ with the following properties:

1) the r.v.s $\{X_{j,k,n}\} \bigcup \{I_{h_n}\}$ $(1 \le k \le n; \ j \in \mathbf{N})$ are independent for every $n \in \mathbf{N}$;

2) $X_{j,k,n} \overset{d}{=} X_j$ $(1 \le k \le n; \ j, n \in \mathbf{N})$.

Let's denote by Y_j the r.v.s determined from $X_{j,k,n}$ by the formula (6). Changing the order of the summation, we obtain

$$\sum_{j=1}^{m} Y_j = \sum_{n=1}^{\infty} \sum_{k=1}^{n} \left(\sum_{j=1}^{m} X_{j,k,n} \right) I_{h_n} .$$

According to 1) and 2), for every integer m

$$\sum_{j=1}^{m} X_{j,k,n} \overset{d}{=} \sum_{j=1}^{m} X_j \quad (1 \le k \le n; \ n \in \mathbf{N}).$$

From what was mentioned above, the series $X_{k,n} = \sum_{j=1}^{\infty} X_{j,k,n}$ are convergent almost surely and $X_{k,n} \overset{d}{=} X$. Since h_n are mutually disjoint sets, this yields that the series $Y = \sum_{k=1}^{\infty} Y_j$ is convergent almost surely and $Y \in \mathcal{L}(\Pi(F))$.

From (7) and (6) $Y_j \ge 0$, which yields $Y \ge Y_j$. Since $\|Y_j\|_{\mathbf{E}} \ge 2^j$ for all integers j, then $Y \notin \mathbf{E}$, which contradicts the condition $\mathbf{E} \in \mathbf{K}$. $\square$

Lemma 2. Let $\mathbf{E} \in \mathbf{K}$ and $\{Z_k\}_{k=1}^{n}$ be mutually disjoint r.v.s with distributions F_k. Let r.v.s $Y_k \in \mathcal{L}(\Pi(F_k))$ be independent. Then

$$e^{-1} \left\| \sum_{k=1}^{n} Z_k \right\|_{\mathbf{E}} \le \left\| \sum_{k=1}^{n} Y_k \right\|_{\mathbf{E}} \le C_1(\mathbf{E}) \left\| \sum_{k=1}^{n} Z_k \right\|_{\mathbf{E}},$$

where $C_1(\mathbf{E})$ is the constant from (5).

Proof: Let F be the distribution of $\sum_{k=1}^{n} Z_k$ and $Y \in \mathcal{L}(\Pi(F))$. One may easy verify, using the assumption that Z_k are mutually disjoint, that

$$\int_{-\infty}^{\infty} (e^{itx} - 1)dF(x) = \sum_{k=1}^{n} \int_{-\infty}^{\infty} (e^{itx} - 1)dF_k(x).$$

Let $g_k(t)$ and $g(t)$ be the characteristic functions of Y_k and Y respectively. The last formula and (4) yield that

$$g(t) = \prod_{k=1}^{n} g_k(t),$$

which implies $Y \stackrel{d}{=} \sum_{k=1}^{n} Y_k$. Lemma 1 gives the needed bounds. $\square$

Lemma 3. *If* $\mathbf{E} \in \mathbf{K}$, *then* $\mathbf{E}'' \in \mathbf{K}$.

Proof: Let $X \in \mathbf{E}''$, the r.v.s $X_{k,n}$ be the same as in (6) and the r.v. Y be defined by this formula. We show that $Y \in \mathbf{E}''$.

Put for an integer m

$$X_{k,m,n} = X_{k,n} I_{\{|X_{k,n}| \leq m\}}, \quad X_m = X I_{\{|X| \leq m\}}.$$

Let the r.v.s Y_m be defined by $X_{k,m,n}$ using the formula (6). Then $X_m \to X$ and $Y_m \to Y$ almost surely. Lemma 1 gives us

$$\|Y_m\|_{\mathbf{E}''} \leq \|Y_m\|_{\mathbf{E}} \leq C_1 \|X_m\|_{\mathbf{E}} = C_1 \|X\|_{\mathbf{E}''}.$$

According to Proposition 5, $Y \in \mathbf{E}''$. $\square$

Now we compare independent and disjoint sums. Let $\{X_k\}_{k=1}^{n}$ be r.v.s which satisfy the condition

$$\sum_{k=1}^{n} P\{X_k \neq 0\} \leq 1. \tag{8}$$

Then there exist mutually disjoint r.v.s $\{\hat{X}_k\}_{k=1}^{n}$ such that

$$\hat{X}_k \stackrel{d}{=} X_k \quad (1 \leq k \leq n).$$

According to Johnson and Schechtman [24], $\{\hat{X}_k\}_{k=1}^{n}$ is said to be a *disjointification* of $\{X_k\}_{k=1}^{n}$.

Lemma 4. *Let* $\mathbf{E} \in \mathbf{K}$ *and independent r.v.s* $\{X_k\}_{k=1}^{n} \subset \mathbf{E}$ *satisfy* (8). *Then*

$$\|\sum_{k=1}^{n} X_k\|_{\mathbf{E}} \leq B(\mathbf{E}) \|\sum_{k=1}^{n} \hat{X}_k\|_{\mathbf{E}}. \tag{9}$$

Proof: We break the proof into three steps.

1. First let's suppose that the r.v.s X_k are symmetric and denote the related distributions by F_k. Let r.v.s $Y_k \in \mathcal{L}(\Pi(F_k))$ be independent. Then Prokhorov's inequality (see section 4) and Proposition 2 yield

$$\|\sum_{k=1}^n X_k\|_{\mathbf{E}} \le 16\|\sum_{k=1}^n Y_k\|_{\mathbf{E}}.$$

Applying Lemma 2, we get the desired estimate with the constant $B = 16C_1(\mathbf{E})$.

2. Let now $\sum_{k=1}^n P\{X_k \ne 0\} \le 1/2$ where X_k are not supposed to be symmetric. Let's denote an independent copy of $\{X_k\}_{k=1}^n$ by $\{Z_k\}_{k=1}^n$ and put $U_k = X_k - Z_k$. From the symmetrization inequality ([20], Ch. 5) $P\{|U_k| > x\} \le 2P\{|X_k| > x/2\}$ for all $x > 0$. Hence the r.v.s $\{U_k\}_{k=1}^n$ satisfy (8) and, therefore, (9).

There are disjointifications $\{\hat{X}_k\}_{k=1}^n$, $\{\hat{Z}_k\}_{k=1}^n$ and $\{\hat{U}_k\}_{k=1}^n$ such that $\hat{U}_k \stackrel{d}{=} \hat{X}_k - \hat{Z}_k$. Since U_k are symmetric, then (9) is fulfilled and we have $\|\sum_{k=1}^n U_k\|_{\mathbf{E}} \le 2B\|\sum_{k=1}^n \hat{X}_k\|_{\mathbf{E}}$.

The relation $U_k = (X_k - EX_k) - (Z_k - EX_k)$ together with Proposition 11 yields

$$\left\|\sum_{k=1}^n U_k\right\|_{\mathbf{E}} \ge C(\mathbf{E}) \left\|\sum_{k=1}^n X_k - E\sum_{k=1}^n X_k\right\|_{\mathbf{E}}$$

$$\ge C(\mathbf{E}) \left(\|\sum_{k=1}^n X_k\|_{\mathbf{E}} - E\left|\sum_{k=1}^n X_k\right|\right).$$

We may suppose that $\|I_\Omega\|_{\mathbf{E}} = 1$. Proposition 1 implies $E|\sum_{k=1}^n X_k| \le \|\sum_{k=1}^n X_k\|_{\mathbf{E}}$ and we get (9) with the constant $B = 1 + 32C_1(\mathbf{E})/C(\mathbf{E})$.

3. Now let's turn to the general case. From (8) $\sum_{k=1}^n P\{|X_k| \ge a\} \le 1/2$ and $\sum_{k=1}^n P\{0 < |X_k| < a\} \le 1/2$ for some positive constant a. Put

$$X_{k,1} = X_k I_{\{|X_k| \ge a\}}, \quad X_{k,2} = X_k - X_{k,1}.$$

Applying the proved inequality to these sequences and taking into account that $|\hat{X}_{k,j}| \le |\hat{X}_k|$ $(j = 1, 2)$, we get (9) with $B(\mathbf{E}) = 2B$, where B was determined above. $\square$

The next Lemma was proved in [24].

Lemma 5. *Let $\{X_k\}_{k=1}^n \subset \mathbf{E}$ be independent r.v.s satisfying (8). If all X_k are non-negative or symmetric, then*

$$\|\sum_{k=1}^n \hat{X}_k\|_{\mathbf{E}} \le 4\|\sum_{k=1}^n X_k\|_{\mathbf{E}}. \tag{10}$$

Proof: We use the following inequality [24]. If $\{X_k\}_{k=1}^n$ are independent non-negative r.v.s, then for all $x > 0$

$$P\left\{\max_{1\le k\le n} X_k \ge x\right\}$$
$$\ge \left(\sum_{k=1}^n P\{X_k \ge x\}\right)\left(1 + \sum_{k=1}^n P\{X_k \ge x\}\right)^{-1}. \tag{11}$$

To prove this we apply the inequalities $1 - x < e^{-x}$ $(0 < x < 1)$ and $1 - e^{-x} > x/(1+x)$ $(x > 0)$, which yield

$$P\left\{\max_{1\le k\le n} X_k \ge x\right\} = 1 - \prod_{k=1}^n (1 - P\{X_k \ge x\})$$
$$\ge 1 - \exp\left(-\sum_{k=1}^n P\{X_k \ge x\}\right)$$
$$\ge \left(\sum_{k=1}^n P\{X_k \ge x\}\right)\left(1 + \sum_{k=1}^n P\{X_k \ge x\}\right)^{-1}.$$

Let X_k be non-negative. Then from (8) and (11)

$$\sum_{k=1}^n P\{X_k \ge x\} \le 2P\left\{\max_{1\le k\le n} X_k \ge x\right\}$$
$$\le 2P\left\{\sum_{k=1}^n X_k \ge x\right\},$$

which implies (10).

Now suppose X_k to be symmetric. Then ([20], Ch. 5)

$$P\left\{\max_{1\le k\le n} |X_k| \ge x\right\} \le 2P\left\{\left|\sum_{k=1}^n X_k\right| \ge x\right\}.$$

From here and (11) $\sum_{k=1}^n P\{|X_k| \ge x\} \le 4P\{|\sum_{k=1}^n X_k| \ge x\}$ and (10) follows. $\square$

The last lemmas give the following result.

Theorem 1. *Let* $\mathbf{E} \in \mathbf{K}$ *and* $\{X_k\}_{k=1}^n \subset \mathbf{E}$ *be independent r.v.s satisfying* (8). *If all the* X_k *are non-negative or symmetric, then*

$$\frac{1}{4}\|\sum_{k=1}^n \hat{X}_k\|_{\mathbf{E}} \le \|\sum_{k=1}^n X_k\|_{\mathbf{E}} \le B(\mathbf{E})\|\sum_{k=1}^n \hat{X}_k\|_{\mathbf{E}}. \tag{12}$$

Now we'll consider some conditions which lead to the Kruglov property. Suppose $X \in \mathcal{L}(F)$ and $Y \in \mathcal{L}(\Pi(F))$. The implication $(Y \in \mathbf{E}) \Rightarrow (X \in \mathbf{E})$ is true for every r.i. space $\mathbf{E}$. Indeed, from (6) $|Y| \geq |X_{1,1}| I_{h_1}$. Since $X_{1,1}$ and I_{h_1} are independent and $X_{1,1} \stackrel{d}{=} X$, then the assumption $(Y \in \mathbf{E})$ yields $(X \in \mathbf{E})$.

Lemma 6. *Let a r.i. space $\mathbf{E}$ be maximal and for every sequence of independent r.v.s $\{X_k\}_{k=1}^{n} \subset \mathbf{E}$ satisfying (8) let the estimate (9) hold. Then $\mathbf{E} \in \mathbf{K}$.*

Proof: Let r.v.s X and I_h be independent and $P(h) = 1/n$. Put $X_n = X I_h$ and let r.v.s $\{X_{k,n}\}_{k=1}^{n}$ be independent and equidistributed with X_n. It is obvious that $\sum_{k=1}^{n} X_{k,n} \stackrel{d}{=} X$. Then, if $X \in \mathbf{E}$, (9) yields the estimate $\|\sum_{k=1}^{n} X_{k,n}\|_{\mathbf{E}} \leq B(\mathbf{E})\|X\|_{\mathbf{E}}$ $(n \in \mathbf{N})$.

Let $f(t)$ be the characteristic function of X. Then the r.v. X_n has the characteristic function $g(t) = n^{-1}f(t) + (1 - n^{-1})$. Hence

$$f_n(t) = \left(n^{-1}(f(t) - 1) + 1\right)^n$$

is the characteristic function of the sum $\sum_{k=1}^{n} X_{k,n}$. Since

$$\lim_{n \to \infty} f_n(t) = \exp(f(t) - 1),$$

these sums are weakly convergent to the r.v. $Y \in \mathcal{L}(\Pi(F))$. The last estimate and Proposition 5 yield $Y \in \mathbf{E}$. $\square$

Theorem 2. *Suppose $\mathbf{E}$ is maximal and $\mathbf{E} \supset L_q(\Omega)$ for some $q < \infty$. Then $\mathbf{E} \in \mathbf{K}$.*

Proof: It was proved by Johnson and Schechtman [24] that the condition $\mathbf{E} \supset L_q(\Omega)$ implies (9) for symmetric r.v.s which satisfy (8). Reasoning as in the proof of Lemma 4, we omit the assumption of symmetry. Lemma 6 gives $\mathbf{E} \in \mathbf{K}$. $\square$

It is unknown whether Lemma 4 and Theorem 2 are true or not for non-maximal r.i. spaces. Here we give another sufficient condition for the Kruglov property.

Theorem 3. *If $\alpha(\mathbf{E}) > 0$, then $\mathbf{E} \in \mathbf{K}$.*

Proof: Using (6), we have for all $x > 0$

$$P\{|X_{k,n} I_{h_n}| \geq x\} = P(h_n)P\{|X_{k,n}| \geq x\}.$$

If $X \in \mathbf{E}$, then from here $\|X_{k,n} I_{h_n}\|_{\mathbf{E}} \leq \gamma_{\mathbf{E}}(P(h_n))\|X\|_{\mathbf{E}}$. The condition $\alpha(\mathbf{E}) > 0$ and Proposition 8 imply $\gamma_{\mathbf{E}}(t) \leq t^{\mu}$ for $0 < t < a$, where μ and a

are positive constants. We have $P(h_n) = 1/n! < a$ for large enough n and $\|X_{k,n}I_{h_n}\|_{\mathbf{E}} \le (1/n!)^{\mu}\|X\|_{\mathbf{E}}$. Since $\sum_{n=1}^{\infty} n(1/n!)^{\mu} < \infty$, then the series (6) is convergent in $\mathbf{E}$ and $Y \in \mathbf{E}$. $\square$

If $\alpha(\mathbf{E}) > 0$, then $\mathbf{E} \supset L_q(\Omega)$ for some $q < \infty$ (see Proposition 9). Hence, if $\mathbf{E}$ is maximal, Theorem 3 follows from Theorem 2. We present the direct proof of Theorem 3 because it is simpler than the proof of Johnson and Schechtman's result.

The condition $\mathbf{E} \in \mathbf{K}$ doesn't imply $\mathbf{E} \supset L_q(\Omega)$ for some $q < \infty$. Let $N(x) = \exp(x) - 1$. Then the Orlicz space $L_N(\Omega)$ has the Kruglov property which follows from Kruglov's Theorem. It is obvious that $L_N(\Omega)$ doesn't contain any $L_q(\Omega)$ $(q < \infty)$.

7. Bases and sequence spaces

Here we give some results about bases in Banach spaces (see, for example, [32] and [38]).

Definition 5. *Let* $\mathbf{B}$ *be a Banach space. A sequence* $\{e_k\}_{k=1}^{\infty} \subset \mathbf{B}$ *is said to be a basis, if each element* $\mathbf{x} \in \mathbf{B}$ *is uniquely represented in the form* $\mathbf{x} = \sum_{k=1}^{\infty} a_k e_k$, *where* $a_k \in \mathbf{R}$ *and the series is convergent in* $\mathbf{B}$.

Definition 6. *Let* $\mathbf{B}$ *be a Banach space. A sequence* $\{\mathbf{x}_k\}_{k=1}^{n} \subset \mathbf{B}$ $(2 \le n \le \infty)$ *is said to be D-unconditional, if*

$$D^{-1}\left\|\sum_{k=1}^{n} a_k \mathbf{x}_k\right\|_{\mathbf{B}} \le \left\|\sum_{k=1}^{n} \epsilon_k a_k \mathbf{x}_k\right\|_{\mathbf{B}} \le D \left\|\sum_{k=1}^{n} a_k \mathbf{x}_k\right\|_{\mathbf{B}}$$

for all $a_k \in \mathbf{R}$ *and* $\epsilon_k = \pm 1$.

Proposition 14. *Let* $\mathbf{E}$ *be a r.i. space and* $\{X_k\}_{k=1}^{\infty} \subset \mathbf{E}$ *be independent r.v.s,* $EX_k = 0$. *Then this sequence is the* $2/C(\mathbf{E})$-*unconditional basis in the subspace* $\text{span}\{X_k\}_{k=1}^{\infty} \subset \mathbf{E}$, *where* $C(\mathbf{E})$ *is the constant from Proposition 11. If the r.v.s* X_k *are symmetric, then this basis is 1-unconditional.*

Proof: According to Proposition 11,

$$\left\|\sum_{k=1}^{m} a_k X_k\right\|_{\mathbf{E}} \le C(\mathbf{E})^{-1} \left\|\sum_{k=1}^{n} a_k X_k\right\|_{\mathbf{E}}$$

for $n > m$ and all $a_k \in \mathbf{R}$. This bound and the Banach criterion (see [38]) imply that $\{X_k\}_{k=1}^{\infty}$ is a basis in the subspace $\text{span}\{X_k\}_{k=1}^{\infty}$.

Let X_k be symmetric. Then the distribution of the sum $\sum_{k=1}^{n} \epsilon_k a_k X_k$ does not depend on $\epsilon_k = \pm 1$. So, the basis $\{X_k\}_{k=1}^{\infty}$ is 1-unconditional.

In the general case we consider an independent copy $\{Y_k\}_{k=1}^{\infty}$ of $\{X_k\}_{k=1}^{\infty}$. Putting $Z_k = X_k - Y_k$ and using Proposition 11, we have

$$C(\mathbf{E}) \left\| \sum_{k=1}^{n} a_k X_k \right\|_{\mathbf{E}} \leq \left\| \sum_{k=1}^{n} a_k Z_k \right\|_{\mathbf{E}} \leq 2 \left\| \sum_{k=1}^{n} a_k X_k \right\|_{\mathbf{E}}. \tag{13}$$

Since the r.v.s Z_k are symmetric, the needed assertion follows. $\square$

The next statement is well known.

Proposition 15. *Let* $\{x_k\}_{k=1}^{n}$, *where* $2 \leq n \leq \infty$, *be a 1-unconditional sequence of elements of a Banach space* $\mathbf{B}$ *and* $|a_k| \leq |b_k|$. *Then*

$$\left\| \sum_{k=1}^{n} a_k x_k \right\|_{\mathbf{B}} \leq \left\| \sum_{k=1}^{n} b_k x_k \right\|_{\mathbf{B}}.$$

Definition 7. *Let* $\{e_k\}_{k=1}^{\infty}$ *be a basis of a Banach space* $\mathbf{B}$. *Let* $x_{j,k} \in \mathbf{R}$ *and* $m_j \nearrow \infty$ *be integers. The sequence*

$$x_j = \sum_{k=m_j+1}^{m_{j+1}} x_{j,k} e_k \tag{14}$$

is said to be a block-basis

Definition 8. *Let* $\mathbf{B}_1$ *and* $\mathbf{B}_2$ *be Banach spaces. The sequences* $\{x_k\}_{k=1}^{\infty} \subset \mathbf{B}_1$ *and* $\{y_k\}_{k=1}^{\infty} \subset \mathbf{B}_2$ *are said to be equivalent (C-equivalent) if there exists a constant* $C > 0$ *such that for all* $a_k \in \mathbf{R}$ *and integers* n

$$C^{-1} \left\| \sum_{k=1}^{n} a_k x_k \right\|_{\mathbf{B}_1} \leq \left\| \sum_{k=1}^{n} a_k y_k \right\|_{\mathbf{B}_2} \leq C \left\| \sum_{k=1}^{n} a_k x_k \right\|_{\mathbf{B}_1}.$$

Let e_k be the infinite real sequence with k-th component equal to 1 and the other equal to 0. We call $\{e_k\}_{k=1}^{\infty}$ the *standard basis*.

Recall that the space l_p $(1 \leq p \leq \infty)$ consists of all real sequences $\mathbf{a} = \{a_k\}_{k=1}^{\infty}$ such that

$$\|\mathbf{a}\|_p = \left(\sum_{k=1}^{\infty} |a_k|^p \right)^{1/p} < \infty \quad (1 \leq p < \infty), \quad \|\mathbf{a}\|_{\infty} = \sup_k |a_k| < \infty.$$

The space c_0 is the subspace of l_{∞} consisting of all real sequences such that $a_k \to 0$.

If $\{a_k\}$ is a real sequence, then $\{a_k^*\}$ is the decreasing rearrangement of the sequence $\{|a_k|\}$. The Lorentz sequence space $l_{r,s}$ consists of all sequences $\mathbf{a} = \{a_k\}_{k=1}^\infty$ for which

$$\|\mathbf{a}\|_{r,s}^* = \left(\sum_{k=1}^\infty (a_k^*)^r \left(k^{r/s} - (k-1)^{r/s} \right) \right)^{1/s} < \infty$$

if $1 \le r$, $s < \infty$ and if $s = \infty$

$$\|\mathbf{a}\|_{r,\infty}^* = \sup_k k^{1/r} a_k^* < \infty .$$

These functionals are quasinorms which are equivalent to norms.

Theorem 4 [54]. *Let a basis $\{\mathbf{x}_k\}_{k=1}^\infty$ of a Banach space $\mathbf{B}$ be equivalent to each of its block-bases. Then $\{\mathbf{x}_k\}_{k=1}^\infty$ is equivalent to the standard basis of the space l_p or c_0.*

Theorem 5 [21]. *Each C-unconditional sequence $\{\mathbf{x}_k\}_{k=1}^\infty$ in a Hilbert space $\mathbf{H}$ is C-equivalent to the orthogonal sequence $\{\mathbf{y}_k\}_{k=1}^\infty$ such that*

$$\|\mathbf{y}_k\|_{\mathbf{H}} = \|\mathbf{x}_k\|_{\mathbf{H}}$$

Definition 9. *A Banach space $\mathcal{E}$ of real sequences is said to be a Banach lattice if the conditions $|x_k| \le |y_k|$ $(k \in \mathbf{N})$ and $\mathbf{y} = \{y_k\}_{k=1}^\infty \in \mathcal{E}$ imply $\mathbf{x} = \{x_k\}_{k=1}^\infty \in \mathcal{E}$ and $\|\mathbf{x}\|_{\mathcal{E}} \le \|\mathbf{y}\|_{\mathcal{E}}$.*

8. Stable distributions

Some well known results about the stable distributions are briefly described here. For details see [20], [22] and [55].

The r.v. X has the symmetric q-stable distribution $(0 < q \le 2)$ if the corresponding characteristic function is determined by the formula

$$f(t) = \exp(-\gamma |t|^q), \tag{16}$$

$\gamma > 0$ is a constant. If $q = 2$, the distribution is said to be normal or gaussian. Putting $\sigma^2 = EX^2$, we have in this case

$$f(t) = \exp\left(\frac{-t^2}{2\sigma^2} \right) . \tag{17}$$

If a r.v. X has the symmetric q-stable distribution and $q < 2$, then there exists the limit

$$\lim_{x \to \infty} x^q P\{|X| \geq x\} = b(q)\gamma \quad (0 < q < 2),$$

where $b(q) > 0$ depends on q only. If X has the symmetric normal distribution and $EX^2 = 1$, then

$$(2\pi)^{-1/2}(x^{-1} - x^{-3})\exp(-x^2/2) \leq P\{|X| \geq x\}$$
$$\leq (2\pi)^{-1/2}x^{-1}\exp(-x^2/2). \tag{19}$$

Let $\{Y_k\}_{k=1}^{\infty}$ be independent r.v.s with the common characteristic function (16). Then for all $a_k \in \mathbf{R}$ and integers n

$$\prod_{k=1}^{n} f(a_k t) = f\left(t \sum_{k=1}^{n} |a_k|^q\right)^{1/q},$$

which yields

$$\sum_{k=1}^{n} a_k Y_k \overset{d}{=} \left(\sum_{k=1}^{n} |a_k|^q\right)^{1/q} Y_1. \tag{20}$$

It should be noted that (20) is the characteristic property of stable distributions (see [55]).

If $0 < p < q < 2$ or $q = 2$ and $p < \infty$, then $\{Y_k\}_{k=1}^{\infty} \subset L_p(\Omega)$. From (20)

$$\left\|\sum_{k=1}^{n} a_k Y_k\right\|_{L_p} = \left(\sum_{k=1}^{n} |a_k|^q\right)^{1/q} \|Y_1\|_{L_p}. \tag{21}$$

CHAPTER II

INEQUALITIES FOR SUMS
OF INDEPENDENT RANDOM VARIABLES
IN REARRANGEMENT INVARIANT SPACES

1. Rosenthal's inequality and a characterization of the spaces L_p

1. Results. Rosenthal proved the following remarkable inequaliy [50]. Let $\{X_k\}_{k=1}^n$ be independent r.v.s such that $EX_k = 0$ and $E|X_k|^p < \infty$ for some $p > 2$. Then

$$\frac{1}{2}\max\left\{\left(\sum_{k=1}^n \|X_k\|_p^p\right)^{1/p}, \left(\sum_{k=1}^n \|X_k\|_2^2\right)^{1/2}\right\} \leq \|\sum_{k=1}^n X_k\|_p$$

$$\leq C(p)\max\left\{\left(\sum_{k=1}^n \|X_k\|_p^p\right)^{1/p}, \left(\sum_{k=1}^n \|X_k\|_2^2\right)^{1/2}\right\},$$

where $C(p)$ depends on p only. In this section we consider problems for which an analog of this estimate holds for r.i. spaces.

Let's turn to explicit statements. Let $\mathcal{E}$ be a Banach lattice of real sequences. We suppose that $\mathcal{E}$ contains the standard basis $\{e_k\}_{k=1}^\infty$. The equality $\mathcal{E}_1 = \mathcal{E}_2$ or $\mathbf{E}_1 = \mathbf{E}_2$ denotes that these Banach lattices, respectively r.i. spaces, coincide as sets. This implies an equivalence of the corresponding norms (see [29]). The intersection of two Banach lattices $\mathcal{E}_1$ and $\mathcal{E}_2$ or r.i. spaces $\mathbf{E}_1$ and $\mathbf{E}_2$ is a Banach lattice, respectively r.i. space, with the norm equal to the maximum of the two initial ones.

There exist r.i. spaces $\mathbf{E}_1 \neq \mathbf{E}_2$ such that

$$a\|X\|_{\mathbf{E}_1} \leq \|X\|_{\mathbf{E}_2} \leq b\|X\|_{\mathbf{E}_1} \tag{1}$$

for all $X \in L_\infty(\Omega)$, where a and b are positive constants independent of X. Indeed, let $\mathbf{E} \neq L_\infty(\Omega)$ be a non-separable r.i. space and $\mathbf{E}_0$ be the closure of $L_\infty(\Omega)$ in $\mathbf{E}$. Then $\mathbf{E} \neq \mathbf{E}_0$ and (1) holds. If $\mathbf{E}_1$ and $\mathbf{E}_2$ are separable, then (1) implies $\mathbf{E}_1 = \mathbf{E}_2$.

Definition 1. *We say that r.i. spaces $\mathbf{E}_1$ and $\mathbf{E}_2$ are essentially different if (1) does not hold for every constant a and b.*

For a r.i. space $\mathbf{E}$, r.v.s $\{X_k\}_{k=1}^n \subset \mathbf{E}$ and a Banach lattice $\mathcal{E}$ we put

$$\mathbf{A}_{\mathbf{E},\mathcal{E}} = \|\{\|X_k\|_{\mathbf{E}}\}_{k=1}^n\|_{\mathcal{E}}. \tag{2}$$

We set for two pairs $(\mathbf{E}_j, \mathcal{E}_j)$ $(j = 1, 2)$ and r.v.s $\{X_k\}_{k=1}^n \subset \mathbf{E}_1 \bigcap \mathbf{E}_2$

$$\mathbf{B}_{\mathbf{E}_1,\mathcal{E}_1,\mathbf{E}_2,\mathcal{E}_2} = \max\left\{\mathbf{A}_{\mathbf{E}_1,\mathcal{E}_1}, \mathbf{A}_{\mathbf{E}_2,\mathcal{E}_2},\right\}. \tag{3}$$

Definition 2. *The expression $\mathbf{E} \in \mathcal{D}(\mathcal{E})$ denotes that for each finite sequence of independent r.v.s $\{X_k\}_{k=1}^n \subset \mathbf{E}$ with mean zero*

$$a_1 \mathbf{A}_{\mathbf{E},\mathcal{E}} \leq \|\sum_{k=1}^n X_k\|_{\mathbf{E}} \leq a_2 \mathbf{A}_{\mathbf{E},\mathcal{E}}, \tag{4}$$

where a_1 and a_2 are positive constants independent of X_k.

It is clear that $L_2(\Omega) \in \mathcal{D}(l_2)$.

Theorem 1. *Let $\mathbf{E} \in \mathbf{K}$ and $\mathbf{E} \in \mathcal{D}(\mathcal{E})$. Then $\mathbf{E} = L_2(\Omega)$ and $\mathcal{E} = l_2$.*

We may consider the analog of (4) where $\mathbf{A}_{\mathbf{E},\mathcal{E}}$ is replaced by $\mathbf{A}_{\mathbf{E}_1,\mathcal{E}}$. But in this connection the result remaines the same because (4) implies (1) and $\mathbf{E}_1 \bigcap \mathbf{E} \in \mathcal{D}(\mathcal{E})$. Theorem 1 yields that $\mathbf{E}_1 \bigcap \mathbf{E} = L_2(\Omega)$. Applying standard reasoning, we get $\mathbf{E}_1 = \mathbf{E} = L_2(\Omega)$.

Definition 3. *We write $\mathbf{E} \in \mathcal{R}(\mathbf{E}_1, \mathcal{E}_1, \mathbf{E}_2, \mathcal{E}_2)$ if for every finite sequence of independent r.v.s $\{X_k\}_{k=1}^n \subset L_\infty(\Omega)$ with mean zero*

$$b_1 \mathbf{B}_{\mathbf{E}_1,\mathcal{E}_1,\mathbf{E}_2,\mathcal{E}_2} \leq \|\sum_{k=1}^n X_k\|_{\mathbf{E}} \leq b_2 \mathbf{B}_{\mathbf{E}_1,\mathcal{E}_1,\mathbf{E}_2,\mathcal{E}_2}, \tag{5}$$

where the constants $b_1, b_2 > 0$ don't depend on X_k.

From Rosenthal's inequality $L_p(\Omega) \in \mathcal{R}(L_p, l_p, L_2, l_2)$ if $2 < p < \infty$. We consider the relation (5) under the assumption that $\mathbf{E}_1$ and $\mathbf{E}_2$ are essentially different and $\mathcal{E}_1 \neq \mathcal{E}_2$. Otherwise we obtain an estimate of the type (4). Indeed, if (1) holds, then there exist positive constants C_1 and C_2 such that

$$C_1 \mathbf{A}_{\mathbf{E}_1,\mathcal{E}_1 \bigcap \mathcal{E}_2} \leq \mathbf{B}_{\mathbf{E}_1,\mathcal{E}_1,\mathbf{E}_2,\mathcal{E}_2} \leq C_2 \mathbf{A}_{\mathbf{E}_1,\mathcal{E}_1 \bigcap \mathcal{E}_2}$$

for all r.v.s $\{X_k\}_{k=1}^n \subset L_\infty(\Omega)$. If $\mathcal{E}_1 = \mathcal{E}_2$, then the same estimate with $\mathbf{A}_{\mathbf{E}_1 \bigcap \mathbf{E}_2,\mathcal{E}_1}$ holds.

Definition 4. *The expression* $(\mathbf{E}_1, \mathcal{E}_1) \subset (\mathbf{E}_2, \mathcal{E}_2)$ *denotes that* $\mathbf{E}_1 \subset \mathbf{E}_2$ *and* $\mathcal{E}_1 \subset \mathcal{E}_2$.

If $(\mathbf{E}_1, \mathcal{E}_1) \subset (\mathbf{E}_2, \mathcal{E}_2)$, then

$$\|X\|_{\mathbf{E}_1} \geq c\|X\|_{\mathbf{E}_2} \quad , \quad \|x\|_{\mathcal{E}_1} \geq d\|x\|_{\mathcal{E}_2}$$

for every $X \in \mathbf{E}_1$ and $x \in \mathcal{E}_1$, where c and d are positive constants (see [29]), which together with (2) and (3) imply

$$c_1 \mathbf{A}_{\mathbf{E}_1, \mathcal{E}_1} \leq \mathbf{B}_{\mathbf{E}_1, \mathcal{E}_1, \mathbf{E}_2, \mathcal{E}_2} \leq c_2 \mathbf{A}_{\mathbf{E}_1, \mathcal{E}_1}.$$

Hence (5) is reduced to (4).

If $\{X_k\}_{k=1}^n \subset L_\infty(\Omega)$, then the expression (3) does not change if $\mathbf{E}_j$ is replaced by $\mathbf{E}_j''$, $j = 1, 2$ (see the section 1.1). From Lemma 1.3, $\mathbf{E} \in \mathbf{K}$ yields $\mathbf{E}'' \in \mathbf{K}$. Moreover, if $\mathbf{E}_1$ and $\mathbf{E}_2$ are essentially different, then $\mathbf{E}_1''$ and $\mathbf{E}_2''$ are the same and vice versa. Hence, without loss of generality, we may suppose $\mathbf{E}_1$ and $\mathbf{E}_2$ to be maximal.

Theorem 2. *Let the r.i. spaces* $\mathbf{E}_1$ *and* $\mathbf{E}_2$ *be maximal and essentially different. Let* $\mathcal{E}_1 \neq \mathcal{E}_2$ *and neither of the couples* $(\mathbf{E}_j, \mathcal{E}_j)$ $(j = 1, 2)$ *be contained in the other. Suppose that* $\mathbf{E}$, $\mathbf{E}_1$ *and* $\mathbf{E}_2$ *have the Kruglov property and* $\mathbf{E} \in \mathcal{R}(E_1, \mathcal{E}_1, \mathbf{E}_2, \mathcal{E}_2)$. *Then there exists* $p \in (2, \infty)$ *such that* $\mathbf{E} = L_p(\Omega)$ *and one of the couples* $(\mathbf{E}_j, \mathcal{E}_j)$ *is* $(L_p(\Omega), l_p)$ *and the other is* $(L_2(\Omega), l_2)$.

The Kruglov property is essential because $L_\infty(\Omega) \in \mathcal{D}(l_1)$, $L_\infty(\Omega) \in \mathcal{R}(L_\infty, l_1, L_2, l_2)$ and $L_\infty(\Omega) \notin \mathbf{K}$.

We break the proofs into several steps.

2. Proof of Theorem 1.

Lemma 1. *If the assumptions of Theorem 1 hold, then* $\mathcal{E} = l_2$.

Proof: Let $\{U_k\}_{k=1}^n$ be a sequence of independent r.v.s with the symmetric Bernoulli distribution. This means that $P\{U_k = -1\} = P\{U_k = 1\} = 1/2$. According to Paley and Zygmund's inequality (see section 1.4)

$$P\left\{ \left| \sum_{k=1}^n a_k U_k \right| \geq \frac{1}{2} \left(\sum_{k=1}^n a_k^2 \right)^{1/2} \right\} \geq \eta,$$

where $\eta > 0$ is independent of a_k and n. Bernstein's inequality implies

$$P\left\{ \left| \sum_{k=1}^n a_k U_k \right| \left(\sum_{k=1}^n a_k^2 \right)^{-1/2} \geq x \right\} \leq 2\exp(-x^2).$$

Since $\mathbf{E} \in \mathbf{K}$, then this space contains a r.v. Z such that $P\{|Z| \geq x\} = \exp(-x^2)$. Indeed, $X \equiv 1 \in \mathbf{E}$ and, therefore, $\mathbf{E}$ contains a r.v. Y with the Poisson distribution. But it is well known that $P\{Y \geq x\} \geq \exp(-x^2)$. Proposition 1.2 implies $Z \in \mathbf{E}$.

So, we have

$$C_1 \left(\sum_{k=1}^{n} a_k^2 \right)^{1/2} \leq \left\| \sum_{k=1}^{n} a_k U_k \right\|_{\mathbf{E}} \leq C_2 \left(\sum_{k=1}^{n} a_k^2 \right)^{1/2}, \tag{6}$$

where $C_1, C_2 > 0$ don't depend on a_k and n.

Without loss of generality we may assume $\|I_\Omega\|_{\mathbf{E}} = 1$. We have

$$\mathbf{A}_{\mathbf{E},\mathcal{E}} = \| \sum_{k=1}^{n} a_k \mathbf{e}_k \|_{\mathcal{E}}$$

for the r.v.s $\{a_k U_k\}_{k=1}^{n}$, where $\{\mathbf{e}_k\}$ is the standard basis. From this, (4) and (6) we have $\mathcal{E} = l_2$. $\square$

Now we show that $\mathbf{E} = L_2(\Omega)$. Theorem 1.1 yields that an estimate of the type (4) is true for mutually disjoint r.v.s. Let's denote the corresponding constants by a_1' and a_2'. First we get some bounds for the fundamental function of $\mathbf{E}$.

Lemma 2. *If the conditions of Theorem 1 are fulfilled, then there exist positive constants d_1 and d_2 it such that*

$$d_1 t^{1/2} \leq \phi_{\mathbf{E}}(t) \leq d_2 t^{1/2} \quad (0 \leq t \leq 1). \tag{7}$$

Proof: For $0 < t \leq 1$ we may choose mutually disjoint sets h_k such that $P(h_k) = t/k \ (1 \leq k \leq n)$. Then

$$\phi_{\mathbf{E}}(t) = \left\| \sum_{k=1}^{n} I_{h_k} \right\|_{\mathbf{E}}.$$

We may assume that $\|\mathbf{x}\|_{\mathcal{E}} = \|\mathbf{x}\|_2$ for every finite x, which yields

$$a_1' \phi_{\mathbf{E}} \left(\frac{t}{n} \right) n^{1/2} \leq \phi_{\mathbf{E}}(t) \leq a_2' \phi_{\mathbf{E}}(\frac{t}{n}) n^{1/2}.$$

Let $0 < r < t \leq 1/2$. Then there is an integer n such that $n < t/r \leq 2n$, which implies $r < t/n \leq 2r \leq 1$. Since the function $\phi_{\mathbf{E}}(t)$ is non-decreasing, then

$$\phi_{\mathbf{E}}(r) \leq \phi_{\mathbf{E}} \left(\frac{t}{n} \right) \leq \phi_{\mathbf{E}}(2r). \tag{8}$$

So

$$a_1' \left(\frac{t}{2r}\right)^{1/2} \phi_{\mathbf{E}}(r) \leq \phi_{\mathbf{E}}(t) \leq a_2' \left(\frac{t}{r}\right)^{1/2} \phi_{\mathbf{E}}(2r).$$

Put $t = 1/2$. Then $r^{-1/2}\phi_{\mathbf{E}}(r) \leq 2\phi_{\mathbf{E}}(1/2)/a_1'$ and $(2r)^{-1/2}\phi_{\mathbf{E}}(2r) \geq \phi_{\mathbf{E}}(1/2)/a_2'$, where $0 < r < 1/2$. Since $\phi_{\mathbf{E}}(1/2) \leq \phi_{\mathbf{E}}(r)r^{-1/2} \leq 2^{1/2}\phi_{\mathbf{E}}(1)$ for $1/2 < r < 1$, we get (7). $\square$

Now we may show that $\mathbf{E} = L_2(\Omega)$. Let

$$X = \sum_{k=1}^{n} a_k I_{h_k}, \tag{9}$$

where h_k are mutually disjoint and $a_k \in \mathbf{R}$. From (2) and (7)

$$d_1\|X\|_2 \leq \mathbf{A}_{\mathbf{E},\varepsilon}\left(\{a_k I_{h_k}\}_{k=1}^{n}\right) \leq d_2\|X\|_2.$$

Since estimates of type (4) are true for mutually disjoint r.v.s, then

$$a_1'd_1\|X\|_2 \leq \|X\|_{\mathbf{E}} \leq a_2'd_2\|X\|_2. \tag{10}$$

Let $Z \in L_2(\Omega)$. We may choose a sequence of r.v.s $\{X_n\}_{n=1}^{\infty}$ of type (9) which is convergent to Z in $L_2(\Omega)$. According to (10), this sequence is fundamental in $\mathbf{E}$. Let's denote the limit of X_n in $\mathbf{E}$ by Y. Since $L_2(\Omega) \subset L_1(\Omega)$ and $\mathbf{E} \subset L_1(\Omega)$, then $Y = Z$ and $L_2(\Omega) \subset \mathbf{E}$. From (10) $\mathbf{E}' = (L_2(\Omega))' = L_2(\Omega)$. Hence $\mathbf{E}'' = L_2(\Omega)$. Since $\mathbf{E} \subset E''$, then $\mathbf{E} = L_2(\Omega)$. $\square$

3. On Banach lattices. Let's turn to the proof of Theorem 2.

Lemma 3. *If the conditions of Theorem 2 hold, then one of the Banach lattices $\mathcal{E}_1$ and $\mathcal{E}_2$ is $l_p = (2 < p < \infty)$ and the other is l_2.*

We break the proof into several steps. Since $\mathbf{E}_1$ and $\mathbf{E}_2$ are essentially different, there is a sequence $\{X_k\}_{k=1}^{\infty} \subset L_\infty(\Omega)$ which is unbounded in one of these spaces (for example in $\mathbf{E}_1$) and bounded in the other. We may choose X_k to be independent and symmetric. Going over to subsequences and norming in $\mathbf{E}_1$, we get the inequalities

$$\|X_k\|_{\mathbf{E}_1} = 1, \quad \|X_k\|_{\mathbf{E}_2} \leq 2^{-k} \quad (k = 1, 2, \dots). \tag{11}$$

First we show, using (11), that the standard basis in $\mathcal{E}_1$ is equivalent to each of its normed block-bases.

Proposition 1. *For every vector* $\mathbf{v} = \{v_j\}_{j=1}^n$ *there exists an integer* $i_0 = i_0(\mathbf{v})$ *such that for all* $i \geq i_0$

$$b_1\|\mathbf{v}\|_{\mathcal{E}_1} \leq \left\|\sum_{j=1}^n v_j X_{i+j}\right\|_{\mathbf{E}} \leq b_2\|\mathbf{v}\|_{\mathcal{E}_1},$$

where b_1 *and* b_2 *are the constants from* (5).

Proof: For the r.v.s $\{v_j X_{i+j}\}_{j=1}^n$ we get, using (2) and (11),

$$\mathbf{A}_{\mathbf{E}_1,\mathcal{E}_2} = \|\mathbf{v}\|_{\mathcal{E}_1} \quad , \quad \mathbf{A}_{\mathbf{E}_2,\mathcal{E}_2} \leq 2^{-i}\|\mathbf{v}\|_{\mathcal{E}_2}.$$

The needed estimates follow from (3) and (5). $\square$

Proposition 2. *The standard basis in* $\mathcal{E}_1$ *is equivalent to each of its normed block-bases.*

Proof: Let the normed block-basis be in the form of (1.14). We have for fixed integer n and $a_j \in \mathbf{R}$

$$\sum_{j=1}^n a_j \mathbf{x}_j = \sum_{j=1}^n \sum_{k=m_j+1}^{m_{j+1}} a_j x_{j,k} \mathbf{e}_k.$$

It follows from Proposition 1 that there exists an integer i_0 such that for $i \geq i_0$

$$b_1 \left\|\sum_{j=1}^n a_j \mathbf{x}_j\right\|_{\mathcal{E}_1} \leq \left\|\sum_{j=1}^n \sum_{k=m_j+1}^{m_{j+1}} a_j x_{j,k} X_{k+i}\right\|_{\mathbf{E}}$$

$$\leq b_2 \left\|\sum_{j=1}^n a_j \mathbf{x}_j\right\|_{\mathcal{E}_1} \tag{12}$$

and for every $j = 1,\dots n$

$$b_1 = b_1\|\mathbf{x}_j\|_{\mathcal{E}_1} \leq \left\|\sum_{k=m_j+1}^{m_{j+1}} x_{j,k} X_{k+i}\right\|_{\mathbf{E}} \leq b_2. \tag{13}$$

The r.v.s

$$Y_{i,j} = \sum_{k=m_j+1}^{m_{j+1}} x_{j,k} X_{k+i} \tag{14}$$

are independent and symmetric for each i. From (2)—(5)

$$b_1\|X\|_{\mathbf{E}_i}\|e_1\|_{\varepsilon_i} \le \|X\|_{\mathbf{E}} \quad (i = 1, 2)$$

for all $X \in L_\infty(\Omega)$. This and (13) imply for all $i \ge i_0$

$$\|Y_{i,j}\|_{\mathbf{E}_1} \le \frac{b_2}{b_1\|e_1\|_{\varepsilon_1}} = c_2 \quad (1 \le j \le n).$$

Using (13) and (14), we obtain for $\{a_j Y_{i,j}\}_{j=1}^n$:

$$b_1 \left\| \sum_{j=1}^n a_j e_j \right\|_{\varepsilon_1} \le \left\| \sum_{j=1}^n \|a_j Y_{i,j}\|_{\mathbf{E}_1} e_j \right\|_{\varepsilon_1}$$

$$= \mathbf{A}_{\mathbf{E}_1,\varepsilon_1} \le c_2 \left\| \sum_{j=1}^n a_j e_j \right\|_{\varepsilon_1}. \tag{15}$$

According to (11)

$$\|Y_{i,j}\|_{\mathbf{E}_2} \le \sum_{k=m_j+1}^{m_{j+1}} |x_{j,k}| \|X_{k+i}\|_{\mathbf{E}_2} \le 2^{-i} \sum_{k=m_j+1}^{m_{j+1}} |x_{j,k}|.$$

Hence $\mathbf{A}_{\mathbf{E}_2,\varepsilon_2} \to 0$ as $i \to \infty$. Applying (3), (5) and (15), we get for sufficiently large i

$$\left\| \sum_{j=1}^n a_j Y_{i,j} \right\|_{\mathbf{E}} \le b_2 c_2 \left\| \sum_{j=1}^n a_j e_j \right\|_{\varepsilon_1}.$$

From the left-hand sides of (5) and (15)

$$\left\| \sum_{j=1}^n a_j Y_{i,j} \right\|_{\mathbf{E}} \ge b_1 \mathbf{A}_{\mathbf{E}_1,\varepsilon_1} \ge b_1^2 \left\| \sum_{j=1}^n a_j e_j \right\|_{\varepsilon_1},$$

where $i \ge i_0$. These estimates, (12) and (14) imply

$$b_1^2 b_2^{-1} \left\| \sum_{j=1}^n a_j e_j \right\|_{\varepsilon_1} \le \left\| \sum_{j=1}^n a_j x_j \right\|_{\varepsilon_1} \le b_1^{-1} b_2 c_2 \left\| \sum_{j=1}^n a_j e_j \right\|_{\varepsilon_1}.$$

So, the block-basis $\{x_j\}_{j=1}^\infty$ is equivalent to the standard basis. $\square$

Proposition 2 and Theorem 1.4 yield that the standard basis in $\mathcal{E}_1$ is equivalent to the standard basis in l_p or c_0, which gives us $\mathcal{E}_1 = l_p$ $(1 \leq p < \infty)$ or $c_0 \subset \mathcal{E}_1 \subset l_\infty$. For finite vectors the norms of c_0 and l_∞ coincide. Therefore, in the last case we may assume that $\mathcal{E}_1 = l_\infty$. Besides we will suppose that $\|\mathbf{x}\|_{\mathcal{E}_1} = \|\mathbf{x}\|_p$ for each $\mathbf{x} \in \mathcal{E}_1$.

From (5) and (6)

$$C_1' \|\mathbf{x}\|_2 \leq \max\{\|\mathbf{x}\|_p, \|\mathbf{x}\|_{\mathcal{E}_2}\} \leq C_2' \|\mathbf{x}\|_2 \tag{16}$$

for every finite real vector $x = \{x_k\}_{k=1}^n$, where C_1' and C_2' are positive constants. Putting $x_k = 1$ $(1 \leq k \leq n)$ and $1/p = 0$ for $p = \infty$, we obtain for every integer n

$$n^{1/p} = \|\mathbf{x}\|_p \leq C_2' \|\mathbf{x}\|_2 = C_2' n^{1/2}.$$

From here $p \geq 2$.

First we consider the case $p \neq 2$.

Proposition 3. *Let (16) hold and $2 < p < \infty$. Then $\mathcal{E}_2 = l_2$.*

Proof: From (16) $\|\mathbf{x}\|_{\mathcal{E}_2} \leq C_2' \|\mathbf{x}\|_2$, which yields $\mathcal{E}_2 \supset l_2$. We show that $\mathcal{E}_2 \subset l_2$. If it is not true, then there is $\mathbf{a} = \{a_k\}_{k=1}^\infty \in \mathcal{E}_2 \backslash l_2$. Hence $\sum_{k=1}^\infty a_k^2 = \infty$ and we may choose a sequence of integers $m_j \nearrow \infty$ such that for the vectors

$$\mathbf{x}_j = \sum_{k=m_j+1}^{m_{j+1}} a_k \mathbf{e}_k$$

the inequality $\|\mathbf{x}_j\|_2 \geq 2^j$ holds. Norming $\mathbf{x}_j$ in l_2, we get finite vectors

$$\mathbf{y}_j = \sum_{k=m_j+1}^{m_{j+1}} b_k \mathbf{e}_k \tag{17}$$

for which

$$\|\mathbf{y}_j\|_2 = 1, \quad \|\mathbf{y}_j\|_{\mathcal{E}_2} \leq 2^{-j} \|\mathbf{x}_j\|_{\mathcal{E}_2}. \tag{18}$$

Let's put $\mathbf{S}_n = \sum_{j=1}^n j^{-1/2} \mathbf{y}_j$ $(n = 1, 2, \dots)$. From (17) and (18)

$$\|\mathbf{S}_n\|_2^2 = \sum_{j=1}^n j^{-1} \to \infty, \quad \sup_n \|\mathbf{S}_n\|_{\mathcal{E}_2} < \infty. \tag{19}$$

Since $p > 2$, then $\|\mathbf{y}_j\|_p \leq \|\mathbf{y}_j\|_2$ and

$$\|\mathbf{S}_n\|_p^p = \sum_{j=1}^n j^{-p/2} \|\mathbf{y}_j\|_p^p \leq \sum_{j=1}^\infty j^{-p/2} < \infty \quad (p < \infty),$$

$$\|\mathbf{S}_n\|_p \leq 1 \quad (p = \infty).$$

Using (16) and (19), we conclude that the sequence $\mathbf{S}_n$ is bounded in l_2. This contradicts (19). So, $\mathcal{E}_2 = l_2$. $\square$

Now we consider the case $\mathcal{E}_1 = l_2$.

Proposition 4. *If $\mathcal{E}_1 = l_2$, then $\mathcal{E}_2 = l_p$ $(p > 2)$.*

Proof: From (16) $\mathcal{E}_1 = l_2 \subset \mathcal{E}_2$. According to the assumptions of Theorem 2, $\mathbf{E}_1$ is not contained in $\mathbf{E}_2$. Since these r.i. spaces are maximal, there exists a sequence $\{Y_k\}_{k=1}^{\infty} \subset L_{\infty}(\Omega)$ which is bounded in $\mathbf{E}_1$ and unbounded in $\mathbf{E}_2$.

Indeed, assume this is not true. It is easy to verify that then $\|Y\|_{\mathbf{E}_2} \leq C\|Y\|_{\mathbf{E}_1}$ for each $Y \in L_{\infty}(\Omega)$ and some constant C. From here ([29], Ch. 1) $(\mathbf{E}_1)_0 \subset (\mathbf{E}_2)_0$, where $\mathbf{E}_0$ denotes the closure of $L_{\infty}(\Omega)$ in $\mathbf{E}$. It is not difficult to deduce that $\mathbf{E}_1 = \mathbf{E}_1'' \subset \mathbf{E}_2'' = \mathbf{E}_2$, which contradicts what was said above.

Reasoning as above, we get $\mathcal{E}_2 = l_p$ $(2 \leq p \leq \infty)$. Since $\mathcal{E}_2 \neq \mathcal{E}_1$ then $p > 2$. $\square$

Lastly we consider the case $\mathcal{E}_2 = l_{\infty}$. The first step is the following assertion.

Proposition 5. *Let the conditions of Theorem 2 hold, $\mathbf{E} = L_2(\Omega)$ and $\mathcal{E}_1 = l_p$ $(2 < p \leq \infty)$. Then $\mathbf{E}_1 \supset \mathbf{E}_2$.*

Proof: Suppose $\mathbf{E}_2$ is not contained in $\mathbf{E}_1$. Then, as above, we get that there are symmetric independent r.v.s $\{X_k\}_{k=1}^{\infty} \subset L_{\infty}(\Omega)$ such that (11) holds. Using (5) for one r.v. X_k, we obtain

$$\inf_k \|X_k\|_{\mathbf{E}} > 0.$$

Since $\mathbf{E} = L_2(\Omega)$, then $\|\sum_{k=1}^{n} X_k\|_{\mathbf{E}} \geq dn^{1/2}$ for all integers n, where $d > 0$ is a constant.

Using (11), (2) and (3), we obtain for the r.v.s $\{X_k\}_{k=1}^{n}$ and n large enough $\mathbf{B}_{\mathbf{E}_1, \mathcal{E}_1, \mathbf{E}_2, \mathcal{E}_2} = n^{1/p}$, where $1/\infty = 0$. From here and (5) $\|\sum_{k=1}^{n} X_k\|_{\mathbf{E}} \leq b_2 n^{1/p}$. Since $2 < p \leq \infty$, this bound contradicts the previous. $\square$

Proposition 6. *Let the conditions of Theorem 2 hold, $\mathcal{E}_1 = l_{\infty}$ and $\mathcal{E}_2 = l_2$. Then $\mathbf{E}_1 \supset \mathbf{E}_2 = \mathbf{E} = L_2(\Omega)$.*

Proof: Using (3), (5) and the arguments of the proof of Lemma 2, we get

$$b_1' \max\left\{\phi_{\mathbf{E}_1}\left(\frac{t}{n}\right), n^{1/2}\phi_{\mathbf{E}_2}\left(\frac{t}{n}\right)\right\} \leq \phi_{\mathbf{E}}(t)$$

$$\leq b_2' \max\left\{\phi_{\mathbf{E}_1}\left(\frac{t}{n}\right), n^{1/2}\phi_{\mathbf{E}_2}\left(\frac{t}{n}\right)\right\}, \tag{20}$$

where $0 < t \leq 1$, $n = 1, 2, \ldots$ and b_1', b_2' are positive constants. Since $\mathbf{E}_1 \in K$, then $\mathbf{E}_1 \neq L_{\infty}(\Omega)$ and $\phi_{\mathbf{E}_1}(r) \to 0$ as $r \to 0$ [48], which gives us

$$b_1' \varlimsup_{n \to \infty} n^{1/2}\phi_{\mathbf{E}_2}\left(\frac{t}{n}\right) \leq \phi_{\mathbf{E}}(t) \leq b_2' \varlimsup_{n \to \infty} n^{1/2}\phi_{\mathbf{E}_2}\left(\frac{t}{n}\right).$$

Putting $t = 1$, we obtain $c_1 n^{1/2} \leq \phi_{E_2}(1/n) \leq c_2 n^{1/2}$, where c_1 and c_2 are positive constants. Using (8) and reasoning as in the proof of Lemma 2, we get

$$d_1 t^{1/2} \leq \phi_{E_2}(t) \leq d_2 t^{1/2} \quad (0 < t \leq 1), \tag{21}$$

where positive d_1 and d_2 are independent of t.

Let X be defined by the formula (9) and $h_{j,k}$, where $1 \leq j \leq m$ and $1 \leq k \leq n$, be mutually disjoint sets such that

$$\bigcup_{j=1}^{m} h_{j,k} = h_k, \quad P(h_{j,k}) = m^{-1} P(h_k).$$

Then

$$X = \sum_{j=1}^{m} \sum_{k=1}^{n} a_k I_{h_{j,k}}.$$

According to (2) and (21), we have for $\{a_k I_{h_{j,k}}\}$

$$d_1 \|X\|_{L_2(\Omega)} \leq \mathbf{A}_{E_2,l_2} \leq d_2 \|X\|_{L_2(\Omega)}.$$

In addition

$$\mathbf{A}_{E_1,l_\infty} = \max_{1 \leq k \leq n} \left\{ |a_k| \phi_{E_1}\left(m^{-1} P(h_k)\right) \right\}.$$

Since $\phi_{E_1}(t) \to 0$ as $t \to 0$, then

$$\mathbf{B}_{E_1,\mathcal{E}_1,E_2,\mathcal{E}_2} = \mathbf{A}_{E_2,l_2}$$

for sufficiently large m. Theorem 1.1 yields that estimates of the type (5) are true for mutually disjoint r.v.s, which implies $v_1 \|X\|_{L_2(\Omega)} \leq \|X\|_E \leq v_2 \|X\|_{L_2(\Omega)}$ for some positive constants v_1 and v_2. So, $E = L_2(\Omega)$.

According to Proposition 5, $E_2 \subset E_1$. Hence $\|X\|_{E_1} \leq C \|X\|_{E_2}$ for each $X \in E_2$, where C is a constant. From (2), (3) and (5) $b_1 \|X\|_{E_2} \leq \|X\|_E \leq b_2(1+C) \|X\|_{E_2}$ for all $X \in L_\infty(\Omega)$. So $E_2 = E = L_2(\Omega)$. $\square$

Proof of Lemma 3: We may assume that $\mathcal{E}_1 = l_p$ $(2 < p \leq \infty)$ and $\mathcal{E}_2 = l_2$. If $p = \infty$, then Proposition 6 yields $E_2 \subset E_1$. Since $l_2 \subset l_\infty$, then $(E_2, \mathcal{E}_2) \subset (E_1, \mathcal{E}_1)$, which contradicts the conditions of Theorem 2. So, $p < \infty$. $\square$

4. On the fundamental functions. Here some inequalities for the fundamental functions of the spaces E, E_1 and E_2 are obtained.

Lemma 4. *Let the conditions of Theorem 2 hold and $\mathcal{E}_1 = l_p$ $(2 < p < \infty)$, $\mathcal{E}_2 = l_2$. Then there are positive constants u, v, u_1, v_1, v_2 such that for $0 < t < 1$*

$$u t^{1/p} \leq \phi_E(t) \leq v t^{1/p}, \tag{22}$$

$$u_1 t^{1/p} \leq \phi_{E_1}(t) \leq v_1 t^{1/p}, \tag{23}$$

$$\phi_{E_2}(t) \leq v_2 t^{1/2}. \tag{24}$$

The proof is based on some auxiliary assertions.

Proposition 7. *There exist positive constants w_1 and w_2 such that for all $0 < r < t < 1/2$*

$$w_1 \max \left\{ \left(\frac{t}{2r} \right)^{1/p} \phi_{\mathbf{E}_1}(r), \left(\frac{t}{2r} \right)^{1/2} \phi_{\mathbf{E}_2}(r) \right\} \leq \phi_{\mathbf{E}}(t)$$

$$\leq w_2 \max \left\{ \left(\frac{t}{r} \right)^{1/p} \phi_{\mathbf{E}_1}(2r), \left(\frac{t}{r} \right)^{1/2} \phi_{\mathbf{E}_2}(2r) \right\}. \tag{25}$$

Proof: For $0 < r < t < 1/2$ there is an integer n such that $r < t/n < 2r < 1$. From here $t/(2r) < n < t/r$. Using (8) and the arguments of the proof of Lemma 2 we obtain (25). $\square$

Proposition 8. *If*

$$\lim_{t \to 0} t^{-1/p} \phi_{\mathbf{E}}(t) = 0,$$

then $\mathbf{E} = L_2(\Omega)$.

Proof: First we obtain some bounds. We may choose $t_n \searrow 0$ such that $t_n^{-1/p} \phi_{\mathbf{E}}(t_n) \to 0$. From (25) for $0 < r < t_n$ and $n = 1, 2, \ldots$

$$r^{-1/p} \phi_{\mathbf{E}_1}(r) \leq \frac{2^{1/p}}{w_1} t_n^{-1/p} \phi_{\mathbf{E}}(t_n),$$

which yields

$$\lim_{r \to 0} r^{-1/p} \phi_{\mathbf{E}_1}(r) = 0. \tag{26}$$

Putting $t = 1/2$ and letting r tend to zero, we obtain from (25)

$$\lim_{r \to 0} r^{-1/2} \phi_{\mathbf{E}_2}(r) > 0.$$

It follows from (25) that $\phi_{\mathbf{E}_2}(r) \leq Cr^{1/2}$ for $0 < r < 1/2$ and some constant C. So, we get (21).

Let $r \to 0$. Then (21), (25) and (26) imply $c_1 t^{1/2} \leq \phi_{\mathbf{E}}(t) \leq c_2 t^{1/2}$ for some positive constants c_1 and c_2. Putting in (25) $t = r$, we get $\phi_{\mathbf{E}_1}(t) \leq (2^{1/p}/w_1) \phi_{\mathbf{E}}(t) \leq wt^{1/2}$, where w is a constant.

Now we may show that $\mathbf{E} = L_2(\Omega)$. Let the r.v. X be defined by the formula (9). For the r.v.s $\{a_k I_{h_k}\}_{k=1}^n$ the following estimates are true:

$$\mathbf{A}_{\mathbf{E}_1, l_p} = \left(\sum_{k=1}^n |a_k|^p \phi_{\mathbf{E}_1} \left(P(h_k) \right)^p \right)^{1/p} \leq \left(\sum_{k=1}^n a_k^2 \phi_{\mathbf{E}_1} \left(P(h_k) \right)^2 \right)^{1/2}$$

$$\leq \left(\sum_{k=1}^n a_k^2 w^2 P(h_k) \right)^{1/2} = w \|X\|_{L_2(\Omega)}.$$

From (21) $d_1\|X\|_{L_2(\Omega)} \leq \mathbf{A}_{\mathbf{E}_2,l_2} \leq d_2\|X\|_{L_2(\Omega)}$. Using (3), we obtain

$$d_1\|X\|_{L_2(\Omega)} \leq \mathbf{B}_{\mathbf{E}_1,l_p,\mathbf{E}_2,l_2} \leq (d_2 + w)\|X\|_{L_2(\Omega)}.$$

Theorem 1.1 (see the section 1.6) and the last bounds imply $\mathbf{E} = L_2(\Omega)$. $\square$

Proof of Lemma 4: First we obtain the upper estimates. Putting in (25) $t = 1/2$, we get (24) and the right-hand side inequality in (23). Hence

$$\phi_{\mathbf{E}}(t) \leq \omega_2 \max\left\{2^{1/p}v_1 t^{1/p}, \ 2^{1/2}v_2 t^{1/2}\right\}.$$

Since $p > 2$, the upper estimate in (22) is proved.

We have

$$\lim_{t \to 0} t^{-1/p}\phi_{\mathbf{E}}(t) > 0. \tag{27}$$

For if not, then according to Proposition 8 $\mathbf{E} = L_2(\Omega)$. Proposition 5 yields $\mathbf{E}_1 \supset \mathbf{E}_2$. Since $l_p \supset l_2$ for $p > 2$, this contradicts the assumptions of Theorem 2.

The relation (27) yields the left-hand side of (22). Putting in (25) $t = r$ and using (22) and (24), we obtain

$$u_1 t^{1/p} \leq w_2 \max\left\{\phi_{\mathbf{E}_1}(2t), \ v_2 2^{1/2}t^{1/2}\right\}.$$

Since $p > 2$, then

$$\lim_{t \to 0} t^{-1/p}\phi_{\mathbf{E}_1}(t) \geq \frac{2^{-1/p}u_1}{w_2} > 0,$$

which implies the left-hand side of (23). $\square$

5. Proof of Theorem 2. We may assume $\mathcal{E}_1 = l_p$ ($2 < p < \infty$) and $\mathcal{E}_2 = l_2$. First we show that $\mathbf{E} = L_p(\Omega)$.

Let X be defined by the formula (9). Applying (23) and (2), we obtain for the r.v.s $\{a_k I_{h_k}\}_{k=1}^n$

$$C_1\|X\|_{L_p(\Omega)} \leq \mathbf{A}_{\mathbf{E}_1,l_p} \leq C_2\|X\|_{L_p(\Omega)},$$

where C_1 and C_2 are positive constants. From (24)

$$\mathbf{A}_{\mathbf{E}_2,l_2} \leq D\|X\|_{L_2(\Omega)} \quad (D = \text{const.}).$$

Since $p > 2$, then according to (3)

$$C_1\|X\|_{L_p(\Omega)} \leq \mathbf{B}_{\mathbf{E}_1,l_p,\mathbf{E}_2,l_2} \leq (C_2 + D)\|X\|_{L_p(\Omega)}.$$

Using Theorem 1.1 and (5), we conclude, as above, that

$$D_1\|X\|_{L_p(\Omega)} \leq \|X\|_{\mathbf{E}} \leq D_2\|X\|_{L_p(\Omega)},$$

where D_1 and D_2 are positive constants. So, $\mathbf{E} = L_p(\Omega)$.

Now we show that $\mathbf{E}_1 \subset \mathbf{E}_2$. If $X \in \mathbf{E}_1$, then for all $x > 0$

$$\|X\|_{\mathbf{E}_1} \geq \|xI_{\{|X| \geq x\}}\|_{\mathbf{E}_1} = x\phi_{\mathbf{E}_1}(P\{|X| \geq x\}). \tag{28}$$

From here and (23) $P\{|X| \geq x\} \leq C(\|X\|_{\mathbf{E}_1})^p x^{-p}$ for some constant C. Using (24) and the condition $p > 2$, we get

$$\int_0^\infty \phi_{\mathbf{E}_2}(P\{|X| \geq x\})dx < \infty.$$

According to Proposition 1.6, $X \in \mathbf{E}_2$. Hence $\mathbf{E}_1 \subset \mathbf{E}_2$ and $\|X\|_{\mathbf{E}_2} \leq B\|X\|_{\mathbf{E}_1}$ for a constant B. Applying (5) to one r.v. X, we get

$$b_1\|X\|_{\mathbf{E}_1} \leq \|X\|_{\mathbf{E}} \leq b_2(B+1)\|X\|_{\mathbf{E}_1}.$$

Therefore $\mathbf{E}_1 = \mathbf{E} = L_p(\Omega)$.

Now we show that $\mathbf{E}_2 = L_2(\Omega)$. Let $X \in L_\infty(\Omega)$ and $\{X_k\}_{k=1}^\infty$ be independent r.v.s equidistributed with X. Put $S_n = n^{-1/2}\sum_{k=1}^n X_k$ and

$$D_n = \max\left\{n^{1/p-1/2}\|X\|_{L_p(\Omega)}, \|X\|_{\mathbf{E}_2}\right\}.$$

From (3) and (5)

$$b_1 D_n \leq \|S_n\|_{L_p(\Omega)} \leq b_2 D_n \quad (n = 1, 2, \dots).$$

Let $\sigma^2 = EX^2$ and let the r.v. Z have the normal distribution with the parameters $(0, 1)$. According to the well known Bernstein's theorem [3]

$$\|S_n\|_{L_p(\Omega)} \to \|\sigma Z\|_{L_p(\Omega)} = \|X\|_{L_2(\Omega)}\|Z\|_{L_p(\Omega)}$$

as $n \to \infty$. Since $p > 2$, then $D_n \to \|X\|_{\mathbf{E}_2}$. Letting $n \to \infty$, we get

$$b_1\|X\|_{\mathbf{E}_2} \leq \|X\|_{L_2(\Omega)}\|Z\|_{L_p(\Omega)} \leq b_2\|X\|_{\mathbf{E}_2}.$$

Hence $\mathbf{E}_2 = L_2(\Omega)$. $\square$

2. Estimates of the von Bahr—Esseen type

1. Introduction and results. Von Bahr and Esseen proved the following inequality [2]. Let $1 < p < 2$ and let $\{X_k\}_{k=1}^n \subset L_p(\Omega)$ be independent r.v.s, $EX_k = 0$ $(1 \leq k \leq n)$. Then

$$\left\|\sum_{k=1}^n X_k\right\|_{L_p} \leq \left(2\sum_{k=1}^n \|X_k\|_{L_p}^p\right)^{1/p}.$$

Definition 5. *We say that a r.i. space* **E** *has the von Bahr and Esseen* p-*property* ($\mathbf{E} \in (\mathbf{BE})_p$) *if*

$$\left\| \sum_{k=1}^{n} X_k \right\|_{\mathbf{E}} \leq B \left(\sum_{k=1}^{n} \|X_k\|_{\mathbf{E}}^{p} \right)^{1/p} \tag{29}$$

for all independent r.v.s $\{X_k\}_{k=1}^{n} \subset \mathbf{E}$, $EX_k = 0$, *where* B *doesn't depend on* X_k.

The estimate (29) may be fulfilled for $p \leq 2$ only. Indeed, let's consider independent r.v.s $\{U_k\}_{k=1}^{\infty}$ with the symmetric Bernoulli distribution. Paley and Zygmund's inequality yields $\|\sum_{k=1}^{n} U_k\|_{\mathbf{E}} \geq C_1 n^{1/2}$, where C_1 is a positive constant. Using (29), we obtain $\|\sum_{k=1}^{n} U_k\|_{\mathbf{E}} \leq C n^{1/p}$ $(n = 1, 2, \ldots)$. So, $p \leq 2$.

If $p = 1$, then (29) turns into the triangle inequality and we suppose in the sequel that $1 < p \leq 2$.

Theorem 3. *Let* $1 < p < 2$ *and* $\mathbf{E} \in \mathbf{K}$. *Then* $\mathbf{E} \in (\mathbf{BE})_p$ *if and only if* $\mathbf{E}$ *satisfies the upper* p-*estimate.*

Theorem 4. *Let* $\mathbf{E} \in \mathbf{K}$. *Then* $\mathbf{E} \in (\mathbf{BE})_2$ *if and only if* $\mathbf{E}$ *satisfies the upper* 2-*estimate and* $\mathbf{E} \subset L_2(\Omega)$.

The condition $\mathbf{E} \in \mathbf{K}$ is essential. Indeed, $L_\infty(\Omega)$ satisfies the upper p-estimate for all $p > 1$ and $L_\infty(\Omega) \notin \mathbf{K}$. One may easily verify that $L_\infty(\Omega) \notin (\mathbf{BE})_p$ if $p > 1$.

2. Some estimates.

Proposition 9. *Suppose* $\mathbf{E}$ *satisfies the upper* p-*estimate. Then* $\phi_{\mathbf{E}}(t) \geq at^{1/p}$ $(0 < t < 1)$, *where* a *is a positive constant.*

Proof: Let $0 < x < 1$. For an integer n we may choose mutually disjoint sets $\{h_k\}_{k=1}^{n}$ such that $P(h_k) = x/n$. We have

$$\phi_{\mathbf{E}}(x) = \left\| \sum_{k=1}^{n} I_{h_k} \right\|_{\mathbf{E}} \leq A \left(\sum_{k=1}^{n} \|I_{h_k}\|_{\mathbf{E}}^{p} \right)^{1/p} = A n^{1/p} \phi_{\mathbf{E}}\left(\frac{x}{n} \right),$$

where A is a constant. For every $t \in (0, 1)$ there is an integer n such that $1/2 \leq tn \leq 1$. Putting $x = tn$, we get

$$\phi_{\mathbf{E}}(t) \geq A^{-1} n^{-1/p} \phi_{\mathbf{E}}(x) \geq A^{-1} \phi_{\mathbf{E}}(1/2) t^{1/p}. \ \square$$

Using (28), we obtain the following statement.

Proposition 10. *Under the assumptions of Proposition 9*

$$P\{|X| \geq x\} \leq bx^{-p} \|X\|_{\mathbf{E}}^{p}$$

for every $X \in \mathbf{E}$ *and all* $x > 0$, *where* b *is a constant.*

Proposition 11. *Suppose* $|X| \leq a$ *and* $1 < p < 2$. *Then*

$$EX^2 \leq \frac{a^{2-p}}{2-p} \left(\|X\|_{p,\infty}^{*} \right)^p .$$

Proof: Put $F(x) = P\{X < x\}$. Since $|X| \leq a$, then

$$EX^2 = \int_{-a}^{a} x^2 dF(x) = - \int_{0}^{a} x^2 d(1 - F(x) + F(-x)).$$

Integrating by parts and taking into account that the term outside of the integral is non-positive, we obtain

$$EX^2 \leq 2 \int_{0}^{a} xP\{|X| \geq x\} dx.$$

According to (1.2), $P\{|X| \geq x\} \leq \left(\|X\|_{p,\infty}^{*} \right)^p x^{-p}$. From here and the previous the needed inequality follows. $\square$

Lemma 5. *Let* $\{U_k\}_{k=1}^{n}$ *be independent r.v.s such that* $|U_k| \leq a$ *and* $EU_k = 0$ $(1 \leq k \leq n)$. *Let a r.i. space* $\mathbf{E}$ *satisfy the upper p-estimate* $(1 < p \leq 2)$ *and if* $p = 2$, *then* $\mathbf{E} \subset L_2(\Omega)$. *Put* $u = \sum_{k=1}^{n} \|U_k\|_{\mathbf{E}}^{p}$. *Then for every* $x > 0$

$$P\left\{ \left| \sum_{k=1}^{n} U_k \right| \geq x \right\} \leq 2 \exp \left(-\frac{x}{2a} \log \left(1 + \frac{\gamma x}{u} \right) \right) \quad (30)$$

where $\gamma > 0$ *depends on* a, p *and* $\mathbf{E}$ *only.*

Proof: Let's put $\sigma^2 = \sum_{k=1}^{n} EU_k^2$. If $p < 2$, then from (1.2) and Proposition 10 $\|X\|_{p,\infty}^{*} \leq b^{1/p} \|X\|_{\mathbf{E}}$ for each $X \in \mathbf{E}$. Proposition 11 yields $\sigma^2 \leq a^{2-p} u/(2-p)$. If $p = 2$, then $\mathbf{E} \subset L_2(\Omega)$. Hence $EX^2 \leq A \|X\|_{\mathbf{E}}^2$ for some constant A, which yields $\sigma^2 \leq Au$. Now we apply Prokhorov's "arcsinh" inequality (see the section 1.4). We have $\mathrm{arcsinh}(t) \geq \log(1+t)$, which yields $\mathrm{arcsinh}\left(ax/(2\sigma^2) \right) \geq \log(1 + \gamma x/u)$, where $\gamma = a/(2A)$ for $p = 2$ and $\gamma = a^{p-1}(2-p)/(2b)$ if $1 < p < 2$. From here (30) follows. $\square$

   3. Proof of Theorem 3. Sufficiency. First we prove (29) for symmetric r.v.s. Let $\{X_k\}_{k=1}^{n} \subset \mathbf{E}$ be independent symmetric r.v.s. We may assume

$$\sum_{k=1}^{n} \|X_k\|_{\mathbf{E}}^{p} = 1. \quad (31)$$

Let's put

$$a = b^{1/p},$$ (31)

where b is the constant from Proposition 10, and

$$U_k = X_k I_{\{|X_k| \le a\}} \quad , \quad V_k = X_k - U_k.$$ (33)

Then

$$\left\| \sum_{k=1}^{n} X_k \right\|_{\mathbf{E}} \le \left\| \sum_{k=1}^{n} U_k \right\|_{\mathbf{E}} + \left\| \sum_{k=1}^{n} V_k \right\|_{\mathbf{E}}.$$ (34)

We estimate each term in the right-hand side.

Let Y be a r.v. with the Poisson distribution. The condition $\mathbf{E} \in \mathbf{K}$ implies $Y \in \mathbf{E}$. It is well known that $P\{Y \ge x\} \ge C \exp(-x \log(1+x))$, where $C > 0$ is a constant. Lemma 5, (31) and (32) imply (30) with $u = 1$. Hence there is a constant $v > 0$ such that for all $x > 0$ the estimate $P\{|\sum_{k=1}^{n} U_k| \ge x\} \le P\{vY \ge x\}$ holds. From here and Proposition 1.2

$$\left\| \sum_{k=1}^{n} U_k \right\|_{\mathbf{E}} \le v \|Y\|_{\mathbf{E}} = D.$$ (35)

Proposition 10 and the relations (31)—(33) imply

$$\sum_{k=1}^{n} P\{V_k \ne 0\} = \sum_{k=1}^{n} P\{|X_k| \ge a\} \le ba \sum_{k=1}^{n} \|X_k\|_{\mathbf{E}}^{p} = 1.$$ (36)

From here and Lemma 1.4,

$$\left\| \sum_{k=1}^{n} V_k \right\|_{\mathbf{E}} \le B(\mathbf{E}) \left\| \sum_{k=1}^{n} \hat{V}_k \right\|_{\mathbf{E}}.$$

Since $\mathbf{E}$ satisfies the upper p-estimate, the term in the right-hand side is not greater than

$$AB(\mathbf{E}) \left(\sum_{k=1}^{n} \|V_k\|_{\mathbf{E}}^{p} \right)^{1/p} \le AB(\mathbf{E}) \left(\sum_{k=1}^{n} \|X_k\|_{\mathbf{E}}^{p} \right)^{1/p} = AB(\mathbf{E}),$$

where A is a constant. So,

$$\left\| \sum_{k=1}^{n} X_k \right\|_{\mathbf{E}} \le AB(\mathbf{E}) + D = H.$$ (37)

Now we consider the general case. Let $\{Y_k\}_{k=1}^n$ be an independent copy of $\{X_k\}_{k=1}^n$ and $Z_k = X_k - Y_k$. Then $\|Z_k\|_\mathbf{E} \leq \|X_k\|_\mathbf{E} + \|Y_k\|_\mathbf{E} = 2\|X_k\|_\mathbf{E}$ and Proposition 1.11 gives us

$$\left\|\sum_{k=1}^n Z_k\right\|_\mathbf{E} \geq C(\mathbf{E}) \left\|\sum_{k=1}^n X_k\right\|_\mathbf{E}.$$

From here (29) follows with $B = 2H/C(\mathbf{E})$. $\square$

4. Proof of Theorem 3. Necessity. Let $\mathbf{E} \in \mathbf{K}$ and $\mathbf{E} \in (\mathbf{BE})_p$. Let $\{X_k\}_{k=1}^n \subset \mathbf{E}$ be mutually disjoint r.v.'s. We may suppose X_k to be symmetric. Applying (29) and Theorem 1.1, we get the desired statement. $\square$

5. Proof of Theorem 4. The sufficiency is proved as above. The inequality (35) follows from the condition $\mathbf{E} \subset L_2(\Omega)$ and the bound (30) for $p = 2$.

Now we turn to the necessity. The upper 2-estimate is proved as above. To show $\mathbf{E} \subset L_2(\Omega)$ we need the next assertions [18].

Lemma 6. *Let* $\{X_k\}_{k=1}^\infty$ *be equidistributed independent r.v.s such that*

$$C \equiv \sup E\left|n^{-1/2}\sum_{k=1}^n X_k\right| < \infty. \tag{38}$$

Then $EX_1 = 0$ *and* $EX_1^2 < \infty$.

Proof: We have $Cn^{1/2} \geq E|\sum_{k=1}^n X_k| \geq |E\sum_{k=1}^n X_k| = n|EX_1|$. From here $EX_1 = 0$.

Let's put $Y_k = X_{2k} - X_{2k-1}$. The r.v.s $\{Y_k\}_{k=1}^\infty$ are symmetric and independent. For $a > 0$ we put

$$Y_{k,a} = Y_k I_{\{|Y_k| < a\}}, \quad U_{k,a} = Y_k - 2Y_{k,a}.$$

From symmetry $U_{k,a} \overset{d}{=} Y_k$ and, therefore,

$$E\left|\sum_{k=1}^n Y_{k,a}\right| = \frac{1}{2}E\left|\sum_{k=1}^n (Y_k - U_{k,a})\right| \leq E\left|\sum_{k=1}^n Y_k\right|.$$

Hence for $\{Y_{k,a}\}_{k=1}^\infty$ the estimate (38) holds with the same constant. Put $\sigma^2(a) = EY_{1,a}^2$ and let Z be a r.v. with the normal distribution and the parameters $(0,1)$. The Central Limit Theorem [35] yields

$$\lim_{n\to\infty} E\left|n^{-1/2}\sum_{k=1}^n Y_{k,a}\right| = \sigma(a)E|Z|.$$

Therefore $\sigma(a)E|Z| \leq C$ for each $a > 0$. Letting $a \to \infty$, we obtain $EY_1^2 < \infty$, which implies $EX_1^2 < \infty$. $\square$

We continue to prove Theorem 4. Let $X \in \mathbf{E}$, $EX = 0$ and $\{X_k\}_{k=1}^\infty$ be independent r.v.s equidistributed with X. Since $\mathbf{E} \in (\mathbf{BE})_2$, the estimate (29) implies

$$\left\| n^{-1/2} \sum_{k=1}^n X_k \right\|_{\mathbf{E}} \leq B \left(n^{-1} \sum_{k=1}^n \|X_k\|_{\mathbf{E}}^2 \right)^{1/2} = B \|X\|_{\mathbf{E}},$$

which together with Proposition 1.1 gives (38). According to Lemma 6, $X \in L_2(\Omega)$ and $\mathbf{E} \subset L_2(\Omega)$. $\square$

3. Upper estimates of the Rosenthal type

1. Introduction and results. Here we'll study the question of the upper estimate in (5). We assume that $\mathbf{E} = \mathbf{E}_1$, $\mathbf{E}_2 = L_2(\Omega)$, $\mathcal{E}_1 = l_p$ and $\mathcal{E}_2 = l_2$, where $2 \leq p < \infty$. For r.v.s $\{X_k\}_{k=1}^n \subset \mathbf{E} \bigcap L_2(\Omega)$ we put

$$\mathbf{B}_{\mathbf{E},p} = \max \left\{ \left(\sum_{k=1}^n \|X_k\|_{\mathbf{E}}^p \right)^{1/p}, \left(\sum_{k=1}^n EX_k^2 \right)^{1/2} \right\}. \tag{39}$$

Definition 6. *We say that a r.i. space $\mathbf{E}$ has the weak Rosenthal p-property $(\mathbf{E} \in (\mathbf{WR})_p)$ if*

$$\left\| \sum_{k=1}^n X_k \right\|_{\mathbf{E}} \leq D\mathbf{B}_{\mathbf{E},p} \tag{40}$$

for all independent r.v.s $\{X_k\}_{k=1}^n \subset \mathbf{E} \bigcap L_2(\Omega)$ with mean zero and some constant D depending on $\mathbf{E}$ and p only.

Theorem 5. *Suppose $\mathbf{E} \in \mathbf{K}$ and $\mathbf{E}$ satisfies the upper p-estimate. Then $\mathbf{E} \in (\mathbf{WR})_p$.*

The condition $\mathbf{E} \in \mathbf{K}$ is essential since the space $L_\infty(\Omega)$ satisfies the upper p-estimate for every $p > 1$ and $L_\infty(\Omega) \notin (\mathbf{WR})_p$ for $p > 2$.

The reverse assertion is not true. Indeed, $L_2(\Omega) \in (\mathbf{WR})_p$ and $L_2(\Omega)$ does not satisfy the upper p-estimate for $p > 2$.

We recall that $\beta(\mathbf{E})$ is the upper Boyd index of $\mathbf{E}$ (see the section 1.2).

Theorem 6. *Let $\mathbf{E} \in \mathbf{K}$, $\beta(\mathbf{E}) < 1/2$ and $\mathbf{E} \in (\mathbf{WR})_p$, where $p > 2$. Then $\mathbf{E}$ satisfies the upper p-estimate.*

Let's consider the case $p = 2$. If $\mathbf{E} \supset L_2(\Omega)$, then $\mathbf{E} \in (\mathbf{WR})_2$. Indeed this assumption yields

$$\|X\|_{\mathbf{E}} \leq C \|X\|_{L_2(\Omega)}$$

for each $X \in L_2(\Omega)$ where C is independent of X. From here (40) follows.

Now we show that if $\mathbf{E} \subset L_2(\Omega)$, then the conditions $\mathbf{E} \in (\mathbf{BE})_2$ and $\mathbf{E} \in (\mathbf{WR})_2$ are equivalent. It is obvious that the first of them implies the second one. Let $\mathbf{E} \in (\mathbf{WR})_2$. Since $\mathbf{E} \subset L_2(\Omega)$, then

$$\|X\|_{L_2(\Omega)} \le D \|X\|_{\mathbf{E}}, \tag{41}$$

where D is a constant. Therefore

$$\mathbf{B}_{\mathbf{E},2} \le \max\{1, D\} \left(\sum_{k=1}^{n} \|X_k\|_{\mathbf{E}}^2 \right)^{1/2}$$

and $\mathbf{E} \in (\mathbf{BE})_2$.

Applying Theorems 3—6 to the Lorentz spaces, which has the Kruglov property (Theorem 1.3), we obtain the folowing result.

Theorem 7. *Let $1 \le p, q \le \infty$ and $r = \min\{p, q\}$. Then*
1) *if $r < 2$, then $L_{p,q}(\Omega) \in (\mathbf{BE})_r$ and $L_{p,q}(\Omega) \notin (\mathbf{BE})_s$ for $s > r$;*
2) *if $r > 2$, then $L_{p,q}(\Omega) \in (\mathbf{WR})_r$ and $L_{p,q}(\Omega) \notin (\mathbf{WR})_s$ for $s > r$;*
3) *if $q = 2 \le p$, then $L_{p,q}(\Omega) \in (\mathbf{BE})_2$ and $L_{p,q}(\Omega) \notin (\mathbf{WR})_s$ for $p > 2$ and $s > 2$;*
4) *if $p = 2 < q$, then $L_{p,q}(\Omega) \notin (\mathbf{BE})_2$ and $L_{p,q}(\Omega) \in (\mathbf{BE})_s$ for $s < 2$.*

2. Proof of Theorem 5. Let $\{X_k\}_{k=1}^n \subset \mathbf{E} \cap L_2(\Omega)$ be independent r.v.s such that

$$\mathbf{B}_{\mathbf{E},p} = 1. \tag{42}$$

We reason the way we did for the proof of Theorem 3. Let r.v.s U_k and V_k be defined by (33). From (39) and (42) $\sigma^2 = \sum_{k=1}^n EU_k^2 \le 1$. Using Prokhorov's "arcsinh" inequality (see the section 1.4), we obtain (30) with $u = 1$, which implies (35). Applying the upper p-estimate, we get the bound for the second term in the right-hand side of (34). So, $\|\sum_{k=1}^n X_k\|_{\mathbf{E}} \le H$, where H depends on $\mathbf{E}$ only, which yields (40). $\square$

3. Proof of Theorem 6. We begin with an auxiliary inequality.

Proposition 12. *Let $\mathbf{E} \subset L_2(\Omega)$, $X \in \mathbf{E}$ and $P\{X \ne 0\} = r$. Then*

$$\|X\|_{L_2(\Omega)} \le Dr^{1/2}\gamma_{\mathbf{E}}\left(\frac{1}{r}\right)\|X\|_{\mathbf{E}}, \tag{43}$$

where $D = D(\mathbf{E})$ is a constant.

Proof: Let $b > 0$. For a r.v. Y denote by $Y^{(b)}$ a r.v. such that for every $x > 0$

$$P\{Y^{(b)} \ge x\} = bP\{Y \ge x\}, \quad P\{Y^{(b)} \le -x\} = bP\{Y \le -x\}.$$

It is easy to see that $Y^{(b)}$ exists if and only if $P\{Y \neq 0\} \leq 1/b$. Since $r \leq 1$, then $X \overset{d}{=} (X^{(1/r)})^{(r)}$. From here and (41) $\|X\|_{L_2(\Omega)} = r^{1/2} \|X^{(1/r)}\|_{L_2(\Omega)} \leq Dr^{1/2} \|X^{(1/r)}\|_{\mathbf{E}} \leq Dr^{1/2} \gamma_{\mathbf{E}} (1/r) \|X\|_{\mathbf{E}} . \ \square$

Let's turn to the proof of Theorem 6. Suppose $\beta(\mathbf{E}) < 1/2$, $\mathbf{E} \in \mathbf{K}$ and $\mathbf{E} \in (\mathbf{WR})_p$, but $\mathbf{E}$ does not satisfy the upper p-estimate. Then for every sequence $a_k \nearrow \infty$ we may choose r.v.s $\{X_{k,n}\}_{k=1}^{m(n)}$ which are mutually disjoint for each $n \in \mathbf{N}$ and such that

$$\left\| \sum_{k=1}^{m(n)} X_{k,n} \right\|_{\mathbf{E}}^p \geq a_n \sum_{k=1}^{m(n)} \|X_{k,n}\|_{\mathbf{E}}^p .$$

Let $0 < r_n < 1$. There are mutually disjoint r.v.s $\{Z_{k,n}\}_{k=1}^{m(n)}$ for which

$$Z_{k,n} \overset{d}{=} X_{k,n}^{(r_n)} \quad (1 \leq k \leq m(n)).$$

It is easy to verify that

$$\sum_{k=1}^{m(n)} Z_{k,n} \overset{d}{=} \left(\sum_{k=1}^{m(n)} X_{k,n} \right)^{(r_n)} .$$

If $0 < r < 1$, then $X \overset{d}{=} (X^{(r)})^{(1/r)}$. Using Proposition 1.7, we get $\|X\|_{\mathbf{E}} \leq \gamma (1/r) \|X^{(r)}\|_{\mathbf{E}} \leq r^{-1} \|X^{(r)}\|_{\mathbf{E}}$, which yields

$$\left\| \sum_{k=1}^{m(n)} Z_{k,n} \right\|_{\mathbf{E}}^p \geq r_n^p \left\| \sum_{k=1}^{m(n)} X_{k,n} \right\|_{\mathbf{E}}^p \geq a_n r_n^p \sum_{k=1}^{m(n)} \|X_{k,n}\|_{\mathbf{E}}^p$$

$$\geq a_n r_n^p \sum_{k=1}^{m(n)} \|Z_{k,n}\|_{\mathbf{E}}^p . \tag{44}$$

We use the inequality $\|X_{k,n}\|_{\mathbf{E}} \geq \|Z_{k,n}\|_{\mathbf{E}}$, which follows from the condition $0 < r_n < 1$.

Since $\beta(\mathbf{E}) < 1/2$, then Proposition 1.9 implies $\mathbf{E} \subset L_2(\Omega)$. According to Proposition 1.8, there are constants $\mu \in (\beta(\mathbf{E}), 1/2)$ and $\nu > 0$ such that $\gamma_{\mathbf{E}}(t) \leq t^\mu$ for $t > \nu$. Choose r_n under the condition

$$r_n^{-1} \geq \max\{1, \nu\}. \tag{45}$$

Then (43) implies

$$\|Z_{k,n}\|_{L_2(\Omega)} \leq Dr_n^{1/2-\mu} \|Z_{k,n}\|_{\mathbf{E}} . \tag{46}$$

Without loss of generality we may assume the r.v.s $Z_{k,n}$ to be symmetric. Let $Y_{k,n}$ be independent r.v.s, $Y_{k,n} \overset{d}{=} Z_{k,n}$. According to (46), we have

$$\mathbf{B}_{\mathbf{E},p} \leq \max \left\{ \left(\sum_{k=1}^{m(n)} \|Y_{k,n}\|_{\mathbf{E}}^p \right)^{1/p}, \; Dr^{1/2-\mu} \left(\sum_{k=1}^{m(n)} \|Y_{k,n}\|_{\mathbf{E}}^2 \right)^{1/2} \right\},$$

where $\mathbf{B}_{\mathbf{E},p}$ determines by the r.v.s $\{X_{k,n}\}_{k=1}^{m(n)}$. Let's put

$$c_n = \sup \frac{\left(\sum_{k=1}^{m(n)} x_k^2 \right)^{1/2}}{\left(\sum_{k=1}^{m(n)} |x_k|^p \right)^{1/p}},$$

where *supremum* is taken over all non-zero vectors on $\mathbf{R}^{m(n)}$. Then

$$\mathbf{B}_{\mathbf{E},p} \leq \max \left\{ 1, \, c_n Dr_n^{1/2-\mu} \right\} \left(\sum_{k=1}^{m(n)} \|Y_{k,n}\|_{\mathbf{E}}^p \right)^{1/p}.$$

Theorem 1.1 and (44) give us

$$\left\| \sum_{k=1}^{m(n)} Y_{k,n} \right\|_{\mathbf{E}}^p \geq \frac{1}{4} \left\| \sum_{k=1}^{m(n)} Z_{k,n} \right\|_{\mathbf{E}}^p \geq \frac{1}{4} a_n r_n^p \sum_{k=1}^{m(n)} \|Z_{k,n}\|_{\mathbf{E}}^p$$

$$= \frac{1}{4} a_n r_n^p \sum_{k=1}^{m(n)} \|Y_{k,n}\|_{\mathbf{E}}^p.$$

Therefore

$$\left\| \sum_{k=1}^{m(n)} Y_{k,n} \right\|_{\mathbf{E}} \geq b_n \mathbf{B}_{\mathbf{E},p}, \tag{47}$$

where

$$b_n = r_n \left(\frac{1}{4} a_n \right)^{1/p} \min \left\{ 1, r_n^{\mu - 1/2} (Dc_n)^{-1} \right\}.$$

Let's choose r_n under the conditions (45) and $r_n^{\mu - 1/2}(Dc_n)^{-1} > 1$. Since $\mu < 1/2$, this is possible. Putting $a_n = 4(n/r_n)^p$, we get $b_n = n$ and (47) gives the bound

$$\left\| \sum_{k=1}^{m(n)} Y_{k,n} \right\|_{\mathbf{E}}^p \geq n \mathbf{B}_{\mathbf{E},p} \quad (n = 1, 2, \dots).$$

This contradicts the condition $\mathbf{E} \in (\mathbf{WR})_p$ and implies the theorem. □

4. Proof of Theorem 7. Since

$$\alpha(L_{p,q}) = \beta(L_{p,q}) = \frac{1}{p} > 0\,, \tag{48}$$

then $L_{p,q}(\Omega) \in \mathbf{K}$. Theorem 3 and Proposition 1.12 imply 1). If $r = \min\{p,q\} > 2$, then from (48)

$$\beta(L_{p,q}) < \frac{1}{2},\,. \tag{49}$$

Using Theorems 5 and 6 and Proposition 1.12, we obtain 2). For $q = 2 \leq p$ we have $L_{p,q}(\Omega) \subset L_2(\Omega)$. Applying Proposition 1.12 and Theorem 4, we get $L_{p,q}(\Omega) \in (\mathbf{BE})_2$. If the strict inequality $q = 2 < p$ holds, then, according to Proposition 1.12, the space $L_{p,q}(\Omega)$ does not satisfy the upper s-estimate for $s > 2$. So, (49) and Theorem 6 yield $L_{p,q}(\Omega) \notin (\mathbf{WR})_s$.

Now let's consider the case $p = 2 < q$. Then the space $L_{p,q}(\Omega)$ is not contained in $L_2(\Omega)$. Theorem 4 gives us $L_{p,q}(\Omega) \notin (\mathbf{BE})_2$. According to Proposition 1.12, $L_{p,q}(\Omega)$ satisfies the upper 2-estimate, which yields the upper s-estimate for $s < 2$. Theorem 3 implies that $L_{p,q}(\Omega) \in (\mathbf{BE})_s$. □

4. Estimates in exponential Orlicz spaces

1. Results. Let's put for $p > 0$

$$N_p(x) = \exp\left(|x|^p\right) - \sum_{k=0}^{[1/p]} \frac{|x|^{kp}}{k!},$$

where $[x]$ denotes the integer part of x. This function is convex and even, $N(0) = 0$. Hence it determines the Orlicz space $L_{N_p}(\Omega)$. In this section we'll consider the question of the von Bahr—Esseen and Rosenthal properties for these spaces.

If $0 < p \leq 1$, then Kruglov's Theorem (see the section 1.6) gives us $L_{N_p}(\Omega) \in \mathbf{K}$. Using the results of the sections 2 and 3 and Proposition 1.13, we conclude that $L_{N_p}(\Omega) \in (\mathbf{WR})_q$ for every $q \geq 2$.

If $p > 1$, then the space $L_{N_p}(\Omega)$ does not have the Kruglov property. Indeed, let F be the distribution of the r.v. $X \equiv 1$. Then $\Pi(F)$ is the Poisson distribution with the parameter $\lambda = 1$. It is easy to verify that $L_{N_p}(\Omega)$ ($p > 1$) does not contain a r.v. $Y \in \mathcal{L}(\Pi(F))$, which yields $L_{N_p}(\Omega) \notin \mathbf{K}$.

Thus, the results of the previous sections cannot be applied to the spaces in question.

We denote the norm on the space $L_{N_p}(\Omega)$ by $\|X\|_{(p)}$. As usual $p' = p/(p-1)$ for $p > 1$. For $\{X_k\}_{k=1}^n \subset L_{N_p}(\Omega)$ we put

$$\mathbf{A}^{(p)} = \left(\sum_{k=1}^{n} \|X_k\|_{(p)}^{p'} \right)^{1/p'}, \tag{50}$$

$$\mathbf{B}^{(p)} = \max \left\{ \mathbf{A}^{(p)}, \left(\sum_{k=1}^{n} EX_k^2 \right)^{1/2} \right\} \tag{51}$$

and

$$\mathbf{H}^{(p)} = \begin{cases} \mathbf{B}^{(p)} & \text{if } 1 < p < 2, \\ \mathbf{A}^{(p)} & \text{if } p \geq 2. \end{cases} \tag{52}$$

Theorem 8. *There exist positive constants $C(p)$ and $D(p)$ such that for every independent r.v.s $\{X_k\}_{k=1}^{n} \subset L_{N_p}(\Omega)$ with mean zero and all $x > 0$*

$$P\left\{ \left| \sum_{k=1}^{n} X_k \right| \geq x \mathbf{H}^{(p)} \right\} \leq C(p) \exp\left(-D(p)x^p\right). \tag{53}$$

From here and Proposition 1.2 the next statement follows.

Theorem 9. *If $1 < p < 2$, then $L_{N_p}(\Omega) \in (\mathbf{WR})_{p'}$ and if $p \geq 2$, then $L_{N_p}(\Omega) \in (\mathbf{BE})_{p'}$.*

Let's consider independent symmetric r.v.s $\{Y_k\}_{k=1}^{\infty}$ such that for every $x > 0$ and $k \in \mathbf{N}$

$$P\{|Y_k| \geq x\} = \exp\left(-x^p\right), \tag{54}$$

where $p > 0$. It is clear that $Y_k \in L_{N_p}(\Omega)$. Put $r(p) = 2$ if $0 < p < 2$, and $r(p) = p'$ if $p \geq 2$.

Theorem 10. *There exist constants $C_j = C_j(p) > 0$ $(j = 1, 2)$ such that for all $a_k \in \mathbf{R}$ and $n \in \mathbf{N}$*

$$C_1 \left(\sum_{k=1}^{n} |a_k|^{r(p)} \right)^{1/r(p)} \leq \left\| \sum_{k=1}^{n} a_k Y_k \right\|_{(p)} \leq C_2 \left(\sum_{k=1}^{n} |a_k|^{r(p)} \right)^{1/r(p)}.$$

According to Theorem 10, $L_{N_p}(\Omega) \notin (\mathbf{BE})_s$ for $p > 2$ and $s > p'$.
To prove these results we need some lemmas.

2. Estimates for characteristic functions. Let X be a r.v. and

$$P\{|X| \geq x\} \leq b \exp\left(-cx^p\right) \tag{55}$$

for every $x > 0$, where $p > 1$ and $b, c > 0$ are constants. Then the corresponding characteristic function is extended to the entire function (see [34]). The following assertion is well known [34].

Lemma 7. *The estimate (55) holds if and only if there are positive constants* β *and* γ *such that*

$$|f(z)| \leq \exp(\beta |z|^{p'}) \tag{56}$$

for all complex z, $|z| \geq \gamma$. *The constants* β *and* γ *are determined by* b, c *and* p *and vice versa.*

If $X \in L_{N_p}(\Omega)$ and $\|X\|_{(p)} = 1$, then one may easily verify that (55) holds with the constants $b = 2$ and $c = 2^{1/p}$. Hence the related characteristic function $f(z)$ is entire.

Let's put for $m \in \mathbf{N}$

$$Q_m(X, z) = \sum_{j=1}^{m} \frac{i^j z^j EX^j}{j!}. \tag{57}$$

Lemma 8. *Let* $X \in L_{N_p}(\Omega)$, $\|X\|_{(p)} = 1$ *and* $m = [p']$. *Let* γ *be the related number from Lemma 7. Then the corresponding characteristic function is represented in the form*

$$f(z) = 1 + Q_m(X, z) + v(z) |z|^{\max\{2, p'\}},$$

where $\sup\{|v(z)| : |z| \leq \gamma, z \in \mathbf{C}\} \leq A(p, \gamma) < \infty$ *and* $A(p, \gamma)$ *depends on* p *and* γ *only.*

Proof: Taylor's formula and the well known equality $EX^k = i^k f^{(k)}(0)$ give us $f(z) = 1 + Q_m(X, z) + T(z)$. The remainder term is represented in the form

$$T(z) = \frac{f^{(m+1)}(u(z))z^{m+1}}{(m+1)!},$$

where $u(z)$ belongs to the segment joining 0 and z. Applying Lemma 7, the formula

$$f^{(m+1)}(u) = i^{m+1} \int_{-\infty}^{\infty} x^{m+1} e^{iux} dF(x),$$

where $F(x) = P\{X < x\}$, and the bound (55) with $b = 2$ and $c = 2^{1/p}$, we obtain the estimate

$$\left| f^{(m+1)}(u(z)) \right| \leq \int_{-\infty}^{\infty} |x|^{m+1} e^{\gamma x} dF(x) \leq B(p, \gamma) < \infty,$$

where $|z| < \gamma$ and $B(p, \gamma)$ depends on p and γ only. Putting

$$v(z) = T(z) |z|^{-\max\{2, p'\}}$$

we obtain the needed representation. $\square$

Let $0 < r_1 < ... < r_n < \infty$. Then

$$\sum_{k=1}^{n} t^{r_k} \le C(t^{r_1} + t^{r_n})$$

for all $t > 0$, where C depends on these exponents only and the inequality

$$\sum_{k=1}^{n} E\,|X|^{r_k} \le C\left(E\,|X|^{r_1} + E\,|X|^{r_n}\right)$$

takes place.

Lemma 9. *Let* $X \in L_{N_p}(\Omega)$, $EX = 0$ *and* $1 < p < 2$. *Then*

$$|f(z)| \le \exp\left(C(p)\left(|z|^2\,EX^2 + |z|^{p'}\,\|X\|_{(p)}^{p'}\right)\right)$$

for every complex z. *If* $p \ge 2$, *then*

$$|f(z)| \le \exp\left(C(p)\min\left\{|z|^2\,\|X\|_{(p)}^2 \;,\; |z|^{p'}\,\|X\|_{(p)}^{p'}\right\}\right).$$

Proof: Assume $\|X\|_{(p)} = 1$. Then (55) holds, where $b = 2$ and $c = 2^{1/p}$. According to Lemma 7 there are positive constants $\beta(p)$ and $\gamma(p)$ such that (56) holds if $|z| \ge \gamma(p)$. Let $1 < p < 2$. Since $EX = 0$, we get from the previous

$$|Q_m(X,z)| \le \sum_{j=2}^{m} \frac{E\,|zX|^j}{j!} + E\,|zX|^{p'} \le C\left(E\,|zX|^2 + E\,|zX|^{p'}\right).$$

There is a constant $D = D(p)$ such that $E\,|Y|^{p'} \le D\,\|X\|_{(p)}^{p'}$ for every $Y \in L_{N_p}(\Omega)$. Hence

$$|Q_m(X,z)| \le C_1\left(|z|^2\,E\,|X|^2 + |z|^{p'}\,\|X\|_{(p)}^{p'}\right).$$

The condition $\|X\|_{(p)} = 1$, Lemma 8 and the inequality $1 + x \le \exp(x)$ imply

$$|f(z)| \le \exp\left(C_2(p)\left(|z|^2\,EX^2 + |z|^{p'}\right)\right),$$

where $|z| \le \gamma(p)$. From here and Lemma 7 the needed bound follows.

If $p \ge 2$, then $m = 1$. Since $EX = 0$, we get $Q_m(X,z) \equiv 0$. From Lemma 8, if $|z| \le \gamma(p)$, then $|f(z)| \le 1 + A(p)\,|z|^2 \le \exp\left(A(p)\,|z|^2\right)$ and, since $p' \le 2$, we obtain from here and (56)

$$|f(z)| \le \exp\left(C(p)\min\{|z|^2 \;,\; |z|^{p'}\}\right).$$

Now we remove the assumption $\|X\|_{(p)} = 1$. Let's put $t = \|X\|_{(p)}$ and $Y = t^{-1}X$. Denoting the characteristic function of Y by $g(z)$, we have $g(z) = f(z/t)$. Using the obtained estimates for $g(z)$, we get the needed estimates for $f(z)$. $\square$

Lemma 10. *Let a r.v. X have the entire characteristic function $f(z)$. Suppose there are positive constants a and b such that $P\{|X| \geq x\} \geq b\exp(-ax^p)$ for $x > b$. Then there exist positive constants α and γ such that $f(-it) \geq \exp\left(\alpha\,|t|^{p'}\right)$ if $|t| \geq \gamma$.*

Proof: We have for $t \in \mathbf{R}$

$$f(-it) = \int_{-\infty}^{\infty} e^{tx}\,dF(x) \geq \int_{\{|x| \geq v|t|^{p'-1}\}} e^{tx}\,dF(x)$$

$$\geq \exp\left(v\,|t|^{p'}\right) P\left\{|X| \geq v\,|t|^{p'-1}\right\},$$

where $F(x) = P\{X < x\}$ and $v > 0$ will be chosen later. From the assumptions of the lemma

$$f(-it) \geq b\exp\left(v\,|t|^{p'} - a\left(v\,|t|^{p'-1}\right)^p\right).$$

Since $p(p'-1) = p'$, then

$$f(-it) \geq \exp\left(v\,|t|^{p'}\left(1 - av^{p-1}\right) + \log(b)\right).$$

Let's choose v under the condition $1 - av^{p-1} > 0$. Then, if $|t|$ is sufficiently large, the needed estimate holds. $\square$

3. Proof of Theorem 8. Let r.v.s $\{X_k\}_{k=1}^n \subset L_{N_p}(\Omega)$ be independent and $EX_k = 0$ $(1 \leq k \leq n)$. We may assume without loss of generality that

$$\mathbf{H}^{(p)} = 1. \tag{58}$$

Let's denote $\lambda_k = \|X\|_{(p)}$ and $Y_k = \lambda_k^{-1} X_k$. Let $f_k(z)$ be the characteristic function of Y_k. Then the sum $S = \sum_{k=1}^n X_k$ has the characteristic function

$$f(z) = \prod_{k=1}^n f_k(\lambda_k z). \tag{59}$$

Suppose $1 < p < 2$. Applying Lemma 9 to every function $f_k(z)$ and using (52) and (58), we obtain

$$|f(z)| \leq \exp\left(C(p)\left(\left(|z|\,\mathbf{H}^{(p)}\right)^2 + \left(|z|\,\mathbf{H}^{(p)}\right)^{p'}\right)\right)$$

$$= \exp\left(C(p)\left(|z|^2 + |z|^{p'}\right)\right).$$

Since $p' > 2$, then $|z|^2 \le |z|^{p'}$ if $|z| \ge 1$. From here and Lemma 7 the needed estimate follows.

Let $p > 2$. According to (59) and Lemma 9

$$|f(z)| \le \exp\left(C(p) \sum_{k=1}^{n} \min\left\{ |\lambda_k z|^2 \ , \ |\lambda_k z|^{p'} \right\} \right).$$

From (58) and (52) $\lambda_k \le 1$. Since $p' \le 2$, then $\lambda_k^2 \le \lambda_k^{p'}$. Hence

$$\min\left\{ |\lambda_k z|^2 \ , \ |\lambda_k z|^{p'} \right\} \le \lambda_k^{p'} \min\left\{ |z|^2 \ , \ |z|^{p'} \right\}.$$

This estimate and (58) imply that if $|z| \ge 1$, then

$$|f(z)| \le \exp\left(C(p) \min\left\{ |z|^2 \ , \ |z|^{p'} \right\} \right) = \exp\left(C(p) |z|^{p'}. \right)$$

Applying Lemma 7, we obtain Theorem 8. □

4. Proof of Theorem 10. Let $0 < p \le 1$. It was mentioned above that $L_{N_p}(\Omega) \in (\mathbf{BE})_2$ in this case. If $1 < p < 2$, then according to Theorem 9, $L_{N_p}(\Omega) \in (\mathbf{WR})_{p'}$. Since $p' > 2$, then $L_{N_p}(\Omega) \in (\mathbf{BE})_2$. So,

$$\left\| \sum_{k=1}^{n} a_k Y_k \right\|_{(p)} \le C_2 \left(\sum_{k=1}^{n} a_k^2 \right)^{1/2}$$

for $0 < p < 2$, where $C_2 = C_2(p)$ is a constant. Since $L_{N_p}(\Omega) \subset L_2(\Omega)$, then

$$\left\| \sum_{k=1}^{n} a_k Y_k \right\|_{(p)} \ge D \left\| \sum_{k=1}^{n} a_k Y_k \right\|_{L_2(\Omega)} = D \left\| Y_1 \right\|_{L_2(\Omega)} \left(\sum_{k=1}^{n} a_k^2 \right)^{1/2},$$

where $D = D(p) > 0$ is a constant. Therefore for $0 < p \le 2$ Theorem 10 is proved.

Let $p > 2$. Theorem 9 implies the upper estimate. To prove the lower one we need the following auxiliary statement.

Lemma 11. *Let the conditions of Theorem 10 be fulfilled. Then for every positive b and c there is a positive $D = D(b, c, p)$ with the following property: if for every $x > 0$*

$$P\left\{ \left| \sum_{k=1}^{n} a_k Y_k \right| \ge x \right\} \le b \exp\left(-cx^p, \right) \tag{60}$$

then

$$\sum_{k=1}^{n} |a_k|^{r(p)} \leq D.$$

Proof: Let $p > 2$. Let's denote the characteristic function of the r.v. Y_1 by $f(z)$. Since Y_1 is symmeric, then

$$f(z) = 1 - \frac{EY_1^2 z^2}{2} + o(|z|^2)$$

as $z \to 0$. Hence $f(-it) \geq \exp\left(\alpha t^2\right)$ for sufficiently small $t \in \mathbf{R}$ and some positive constant α. It is easy to verify that from (54) the strong inequality $f(-it) > 1$ follows for each real non-zero t. Applying Lemma 10, we obtain the estimate

$$f(-it) \geq \exp\left(\delta \min\left\{t^2, |t|^{p'}\right\}\right),$$

which holds for all $t \in \mathbf{R}$ and a positive constant δ.

Suppose (60) holds. Let's denote the characteristic function of the sum $S = \sum_{k=1}^{n} a_k Y_k$ by $g(z)$. We have

$$g(-it) = \prod_{k=1}^{n} f(-ia_k t) \geq \exp\left(\delta \sum_{k=1}^{n} \min\left\{(a_k t)^2, |a_k t| p'\right\}.\right)$$

Lemma 7 and (60) imply that there are positive constants β and γ depending on b and c only, such that

$$|g(z)| \leq \exp\left(\beta |z|^{p'}\right)$$

if $|z| \geq \gamma$. From the last relations

$$\delta \sum_{k=1}^{n} \min\left\{(a_k t)^2, |a_k t|^{p'}\right\} \leq \beta |t|^{p'}$$

if $|t| \geq \gamma$. Since $r(p) = p' < 2$, then

$$\sum_{k=1}^{n} |a_k|^{r(p)} \leq \frac{\beta}{\delta}$$

and in the case $p > 2$ the proof is complete.

If $0 < p \leq 2$, then (60) gives us

$$\sum_{k=1}^{n} a_k^2 EY_1^2 = E\left(\sum_{k=1}^{n} a_k Y_k\right)^2 \leq C(b,c) < \infty$$

and Lemma follows. □

We continue to prove Theorem 10. Suppose that for $p > 2$ the lower estimate is not true. Then there are real numbers $\{a_{k,j}\}_{k=1}^{n(j)}(j = 1, 2, \ldots)$ such that

$$\sum_{k=1}^{n(j)} |a_{k,j}|^{r(p)} = 1 \quad , \quad \left\| \sum_{k=1}^{n(j)} a_{k,j} Y_k \right\|_{(p)} \leq 2^{-j}. \tag{61}$$

Put $m(0) = 0$ and $m(j) = n(1) + \cdots + n(j)$ and consider the sums

$$S_i = \sum_{j=1}^{i} \sum_{k=1}^{n(j)} a_{k,j} Y_{m(j-1)+k}. \tag{62}$$

According to (61), $\|S_i\|_{(p)} \leq 1$ for all $i \in \mathbf{N}$, which implies the estimate (55) for S_i with the constants $b = 2$ and $c = 2^{1/p}$. Applying Lemma 11, we obtain

$$\sum_{j=1}^{i} \sum_{k=1}^{n(j)} |a_{k,j}|^{r(p)} \leq D,$$

where D does not depend on i. But according to (61), the left-hand side is equal to i. Hence the last inequality cannot be true for all i. This contradiction proves Theorem 10. □

CHAPTER III

LINEAR COMBINATIONS
OF INDEPENDENT RANDOM VARIABLES
IN REARRANGEMENT INVARIANT SPACES

1. l_q-estimates $(q \neq 2)$

1. Introduction and results. Throughout this chapter $\{X_k\}_{k=1}^{\infty}$ is a sequence of independent identically distributed random variables (i.i.d.r.v.s), $\mathbf{E}$ is a r.i. space and q is a fixed positive number. We suppose that $\{X_k\}_{k=1}^{\infty} \subset \mathbf{E}$ and consider estimates of the types

$$\left\| \sum_{k=1}^{n} a_k X_k \right\|_{\mathbf{E}} \leq C \left\| \mathbf{a} \right\|_q$$

and

$$\left\| \sum_{k=1}^{n} a_k X_k \right\|_{\mathbf{E}} \geq B \left\| \mathbf{a} \right\|_q ,$$

where $\mathbf{a} = \{a_k\}_{k=1}^{n}$ and B and C are positive constants independent of n and a_k. We call these estimates the upper and lower l_q-estimates respectively.

Theorem 1. *Suppose there exist positive constants C_1 and C_2 such that for every $n \in \mathbf{N}$*

$$C_1 n^{1/q} \leq \left\| \sum_{k=1}^{n} X_k \right\|_{\mathbf{E}} \leq C_2 n^{1/q}. \tag{1}$$

Then $1 \leq q \leq 2$. If $q > 1$, then $EX_1 = 0$.

Let Z_q be a symmetric r.v. such that for all $x > 0$

$$P\{|Z_q| \geq x\} = \min\{x^{-q}, 1\}. \tag{2}$$

Theorem 2. *Let $1 < q < 2$, $\mathbf{E} \supset L_{q,\infty}(\Omega)$ and*

$$\lim_{x \to \infty} \left\| Z_q I_{\{|Z_q| \geq x\}} \right\|_{\mathbf{E}} = 0. \tag{3}$$

Suppose (1) is fulfilled. Then $EX_1 = 0$ and there exist positive constants a, b and c such that for $x > c$

$$ax^{-q} \leq P\{|X_1| \geq x\} \leq bx^{-q}. \tag{4}$$

In this connection the upper estimate in (1) implies the same estimate in (4) and the condition $EX_1 = 0$.

Theorem 3. *Suppose* $1 < q < 2$, $\mathbf{E} \supset L_{q,\infty}(\Omega)$, $EX_1 = 0$ *and (4) holds. Then for all* $a_k \in \mathbf{R}$ *and integers* n

$$D_1 \left(\sum_{k=1}^{n} |a_k|^q \right)^{1/q} \le \left\| \sum_{k=1}^{n} a_k X_k \right\|_{\mathbf{E}} \le D_2 \left(\sum_{k=1}^{n} |a_k|^q \right)^{1/q} \tag{5}$$

where D_1 *and* D_2 *are positive constants depending on* $\mathbf{E}$ *and the constants from (4) only. In addition, the right-hand side of (5) follows from the right-hand side of (4).*

The next result shows that the condition (3) is essential. Indeed, for the space $L_{q,\infty}(\Omega)$ this condition doesn't hold.

Theorem 4. *Let* $1 < q < 2$. *There are i.i.d.r.v.s* $\{X_k\}_{k=1}^{\infty} \subset L_{q,\infty}(\Omega)$ *such that (5) holds and the lower estimate in (4) doesn't hold.*

The condition $\mathbf{E} \supset L_{q,\infty}(\Omega)$ is also essential. Consider the Orlicz space $L_{N_p}(\Omega)$ and i.i.d.r.v.s $\{Y_k\}_{k=1}^{\infty}$ for which (2.54) is fulfilled. If $p > 2$, then according to Theorem 2.10, the relation (5) holds, where $\mathbf{E} = L_{N_p}(\Omega)$ and $q = p'$. But the lower estimate in (4) does not hold. It is clear that $L_{N_p}(\Omega)$ does not contain the space $L_{q,\infty}(\Omega)$.

One may easy verify that for every sequence of i.i.d.r.v.s $\{X_k\}_{k=1}^{\infty} \subset L_{\infty}(\Omega)$ such that $EX_1 = 0$, the inequality (5) holds, where $q = 1$. Conversely, if a r.i. space $\mathbf{E}$ has such a property, then $\mathbf{E} = L_{\infty}(\Omega)$ (see [**48**]).

Now we consider the spaces $L_p(\Omega)$ for $0 < p < \infty$. If $p < 1$, the space $L_p(\Omega)$ is not a normed space and therefore Theorem 1 cannot be applied.

Theorem 5. *Let* $p, q > 0$, $q \ne 2$ *and let* $\{X_k\}_{k=1}^{\infty}$ *be i.i.d.r.v.s. Suppose for* $\mathbf{E} = L_p(\Omega)$ *(1) holds and if* $q = 1$

$$\inf_{n} \left\| \sum_{k=1}^{n} n^{-1/q} (-1)^k X_k \right\|_{L_p(\Omega)} > 0.$$

Then
1) $0 < p < q < 2$;
2) *(4) holds;*
3) *if* $q > 1$, *then* $EX_1 = 0$;
4) *if* $q = 1$, *then*

$$\sup_{a>0} \left| EX_1 I_{\{|X_1|<a\}} \right| < \infty.$$

In addition, the upper estimate in (1) and the condition 1) imply the upper estimate in (4).

The supplementary condition to (1) is essential. Indeed, let $X_k \equiv 1$, $q = 1$ and $p > 0$. Then (1) holds, but the lower estimate in (4) is not true.

Theorem 6. *Let the conditions 1)–4) of Theorem 5 be fulfilled. Then for* $\mathbf{E} = L_p(\Omega)$ *the estimate (5) holds. In this connection the upper estimate in (4) implies the same estimate in (5).*

It should be noted that under the assumptions $\mathbf{E} = L_p(\Omega)$ and $0 < p < q < 2$ the equivalence of the upper estimate in (1) and (4) was proved in [**18**].

2. Proof of Theorem 1. First we prove an auxiliary result.

Lemma 1. *Let* $\mathbf{E}$ *be a r.i. space,* $\{X_k\}_{k=1}^{\infty} \subset \mathbf{E}$ *be i.i.d.r.v.s and* $X_1 \neq 0$. *Then there is a constant* $C > 0$, *depending on* $\mathbf{E}$ *and the distribution of the r.v.* X_1 *only, such that*

$$\left\| \sum_{k=1}^{n} a_k X_k \right\|_{\mathbf{E}} \geq C \left(\sum_{k=1}^{n} a_k^2 \right)^{1/2} \tag{6}$$

for all $a_k \in \mathbf{R}$ *and integers* n.

Proof: Put $Y_k = X_{2k} - X_{2k-1}$ and for $b > 0$

$$Y_{k,b} = Y_k I_{\{|Y_k| < b\}} , \quad U_{k,b} = 2Y_{k,b} - Y_k.$$

In view of symmetry $U_{k,b} \overset{d}{=} Y_k$. Since $(U_{k,b} + Y_k)/2 = Y_{k,b}$, then

$$\left\| \sum_{k=1}^{n} a_k Y_{k,b} \right\|_{\mathbf{E}} \leq \frac{1}{2} \left(\left\| \sum_{k=1}^{n} a_k Y_k \right\|_{\mathbf{E}} + \left\| \sum_{k=1}^{n} a_k U_{k,b} \right\|_{\mathbf{E}} \right) = \left\| \sum_{k=1}^{n} a_k Y_k \right\|_{\mathbf{E}}.$$

Choosing $b > 0$ so that $Y_{k,b} \neq 0$ and applying Paley and Zygmund's inequality (see section 1.4), we obtain

$$P \left\{ \left| \sum_{k=1}^{n} Y_{k,b} \right| \geq \frac{1}{2} \left(\sum_{k=1}^{n} a_k^2 \right)^{1/2} \right\} \geq \eta$$

where $\eta > 0$ is independent of n and a_k. From here

$$\left\| \sum_{k=1}^{n} a_k Y_{k,b} \right\|_{\mathbf{E}} \geq D \left(\sum_{k=1}^{n} a_k^2 \right)^{1/2},$$

where $D > 0$ is a constant. Taking into account the above, we obtain (6) for $\{Y_k\}_{k=1}^{\infty}$. Since the r.v.s $\{X_k\}_{k=1}^{\infty}$ are identically distributed, then

$$\left\| \sum_{k=1}^{n} a_k Y_k \right\|_{\mathbf{E}} \leq 2 \left\| \sum_{k=1}^{n} a_k X_k \right\|_{\mathbf{E}},$$

which implies the needed estimate. $\square$

Let's turn to the proof of Theorem 1. According to (1)

$$C_1 n^{1/q} \le \left\| \sum_{k=1}^{n} X_k \right\|_{\mathbf{E}} \le \sum_{k=1}^{n} \|X_k\|_{\mathbf{E}} = n \, \|X_1\|_{\mathbf{E}}$$

for every integer n. Therefore $q \ge 1$. Using (1) and (6) we get

$$Cn^{1/2} \le \left\| \sum_{k=1}^{n} X_k \right\|_{\mathbf{E}} \le C_2 n^{1/q},$$

which implies $q \le 2$.

Proposition 1.1 and (1) give us

$$C_2 n^{1/q} \ge \left\| \sum_{k=1}^{n} X_k \right\|_{\mathbf{E}} \ge A \left\| \sum_{k=1}^{n} X_k \right\|_{L_1(\Omega)} \ge A \left| E \sum_{k=1}^{n} X_k \right| = An \, |EX_1|,$$

where $A > 0$ depends on $\mathbf{E}$ only. If $q > 1$, it follows that $EX_1 = 0$. $\square$

3. Some estimates for characteristic functions. To prove the results mentioned above we use characteristic functions. Here we find out how (4) affects the behavior of the related characteristic function near zero.

Lemma 2. *Let $f(x)$ be the characteristic function corresponding to the r.v. X and $0 < q < 2$. The condition (4) holds if and only if there are positive constants α, β and γ such that*

$$\alpha \, |t|^q \le 1 - \mathrm{Re}f(t) \le \beta \, |t|^q \tag{7}$$

for $|t| < \gamma$. The constants α, β and γ are determined by a, b and c and vice versa. The upper estimate (4) is equivalent to the same estimate in (7).

First we prove the following auxiliary bound.

Proposition 1. *Let the upper estimate in (4) hold and denote $F(x) = P\{X < x\}$. Then for every $r > q$ and $t > 0$*

$$\int_{-1/t}^{1/t} |x|^r \, dF(x) \le \frac{r}{(r-q)} \max \{b, c^q\} \, t^{q-r}.$$

Proof: It is obvious that

$$\int_{-1/t}^{1/t} |x|^r \, dF(x) = -\int_{0}^{1/t} x^r \, d \left(1 - F(x) + F(-x)\right). \tag{8}$$

We integrate by parts. Outside the integral we have

$$-\frac{1}{t^r}\left(1 - F\left(\frac{1}{t}\right) + F\left(-\frac{1}{t}\right)\right) < 0$$

for all $t > 0$. Hence

$$\int_{-1/t}^{1/t} |x|^r \, dF(x) \leq r \int_0^{1/t} x^{r-1} \left(1 - F(x) + F(-x)\right) dx.$$

According to (4),

$$1 - F(x) + F(-x) \leq P\{|X| \geq x\} \leq bx^{-q}$$

for $x > c$. If $0 < x < c$, then $1 - F(x) + F(-x) \leq 1 \leq c^q x^{-q}$. The last inequalities imply the needed bound. $\square$

Proof of Lemma 2. Implication (4) $\Rightarrow$ (7): We have

$$1 - \mathrm{Re}f(t) = \int_{-\infty}^{\infty} (1 - \cos(tx)) \, dF(x), \qquad (9)$$

where $F(x)$ is the same as above. Since $0 \leq 1 - \cos(x) \leq \min\{2, x^2\}$ then for $t \neq 0$

$$1 - \mathrm{Re}f(t) \leq \frac{t^2}{2} \int_{|x| \leq 1/t} x^2 dF(x) + 2 \int_{|x| > 1/t} x^2 dF(x).$$

We estimate the first integral using Proposition 1 with $r = 2$. The second integral equals $2P\{|X| \geq |t|^{-1}\}$. According to (4) it is not greater than $2b|t|^q$ if $|t| \geq c$. Therefore, if $|t| \leq 1/c = \gamma$, then the upper estimate (7) holds, where

$$\beta = 2b + \frac{\max\{b, c^q\}}{2 - q}. \qquad (10)$$

Our arguments show that this estimate follows from the upper one in (4).

Let's turn to the lower estimate. Applying (9) and the well known inequality $1 - \cos(x) \geq x^2/3$ ($|x| \leq 1$), we obtain

$$1 - \mathrm{Re}f(t) \geq \frac{t^2}{3} \int_{\delta/|t| \leq |x| \leq 1/|t|} x^2 dF(x) \qquad (11)$$

for all $t \neq 0$ and every $\delta \in (0, 1)$. From here

$$1 - \operatorname{Re} f(t) \geq \frac{t^2 \delta^2}{3 \, t^2} \left(F\left(\frac{1}{|t|} + 0\right) - F\left(\frac{\delta}{|t|}\right) \right)$$

$$+ \frac{t^2 \delta^2}{3 \, t^2} \left(F\left(-\frac{\delta}{|t|} + 0\right) - F\left(-\frac{1}{|t|}\right) \right)$$

$$= \frac{\delta^2}{3} \left(P\left\{ |X| \geq \frac{\delta}{|t|} \right\} - P\left\{ |X| \geq \frac{1}{|t|} \right\} \right).$$

Using the estimates (4), we get for $0 < |t| < \delta/c$

$$1 - \operatorname{Re} f(t) \geq \frac{\delta^2}{3} \left(a \left(\frac{\delta}{|t|}\right)^{-q} - b \left(\frac{1}{|t|}\right)^{-q} \right) = \frac{\delta^2}{3} \left(a\delta^{-q} - b \right) |t|^q.$$

Choosing $\delta \in (0, 1)$ under the condition $a\delta^{-q} - b > 0$ and putting

$$\alpha = \frac{\delta^2 (a\delta^{-q} - b)}{3} \quad , \quad \gamma = \frac{\delta}{c},$$

we get the lower estimate in (4).

Implication $(7) \Rightarrow (4)$: First we prove the upper estimate. The well known inequality [35]

$$P\left\{ |X| \geq \frac{1}{t} \right\} \leq \frac{7}{t} \int_0^t (1 - \operatorname{Re} f(x)) \, dx \quad (t > 0)$$

and the upper estimate in (7) yield that

$$P\left\{ |X| \geq \frac{1}{t} \right\} \leq \frac{7\beta t^q}{1 + q}$$

for $0 < t < \gamma$. Putting $x = 1/t$, $c = 1/\gamma$ and $b = 7\beta(1 + q)$, we obtain the right-hand side inequality in (4).

Let's turn to the lower estimate. Putting in (11) $\delta = 0$ and taking into account (7), we get for $0 < r < \gamma$

$$\int_{-1/r}^{1/r} x^2 dF(x) \leq 3\beta r^{q-2}.$$

Let $t = \delta r$ and $0 < \delta < 1$. From (7), (9) and the last inequality

$$\alpha t^q \leq 1 - \operatorname{Re} f(t) \leq \frac{t^2}{2} \int_{-\delta/t}^{\delta/t} x^2 dF(x) + 2 \int_{|x| \geq \delta/t} dF(x)$$

$$\leq \frac{3\beta}{2} \left(\frac{\delta}{t}\right)^{2-q} t^2 + 2P\left\{ |X| \geq \frac{\delta}{t} \right\},$$

where $0 < t < \gamma\delta$. So,

$$P\left\{|X| \geq \frac{\delta}{t}\right\} \geq \left(\alpha - \frac{3\beta\delta^{2-q}}{2}\right)\frac{t^q}{2}$$

for such t. Let's denote

$$\delta = \left(\frac{\alpha}{3\beta}\right)^{1/(2-q)} \quad , \quad a = \frac{\alpha\delta^q}{4}.$$

Since $0 < q < 2$ and $0 < \alpha < \beta$, then $0 < \delta < 1$. Let $x = \delta/t$. The last relations imply that if $x > c = 1/\gamma$, then the lower estimate in (4) holds. $\square$

Similar results were proved in [4] and [5].

Now we consider estimates for the imaginary part of the characteristic function. We study conditions under which there are positive constants μ and ν such that

$$|\mathrm{Im}f(t)| \leq \mu\,|t|^q \tag{12}$$

if $|t| \leq \nu$.

Lemma 3. *Let $f(t)$ be the characteristic function of the r.v. X and $0 < q < 2$. Suppose the upper estimate in (4) holds. Then the following hold:*

1) if $0 < q < 1$, then (12) is true;

2) if $q = 1$, then (12) holds if and only if

$$\sup_{a>0}\left|EXI_{\{|X|\leq a\}}\right| < \infty;$$

3) if $1 < q < 2$, then (12) holds if and only if $EX = 0$.

Proof: We have

$$\mathrm{Im}f(t) = \int_{-\infty}^{\infty} \sin(tx)dF(x), \tag{13}$$

where $F(x)$ is the corresponding distribution function. Let's put

$$J_1(t) = \int_{-1/|t|}^{1/|t|} (\sin(tx) - tx)\,dF(x) \quad , \quad J_2(t) = t\int_{-1/|t|}^{1/|t|} x dF(x),$$

$$J_3(t) = \int_{|x|\geq 1/|t|} \sin(tx)dF(x),$$

where $t \neq 0$. Then $\mathrm{Im}f(t) = J_1(t) + J_2(t) + J_3(t)$.

It is well known that $|\sin(x) - x| \leq |x|^3/6$ if $|x| \leq 1$. Using Proposition 1, we get

$$|J_1(t)| \leq \frac{\max\,(b,c^q)\,|t|^q}{6 - 2q}.$$

As $|\sin(x)| \leq 1$, the upper estimate in (4) implies the inequality $|J_3(t)| \leq b\,|t|^q$ for $|t| \leq 1/c$. It follows from the last relations that (12) holds if and only if

$$J_2(t) = O(|t|^q) \quad (t \to 0). \tag{14}$$

Let $0 < q < 1$. Applying Proposition 1 for $r = 1 > q$, we get (14). From the definition of $J_2(t)$, for $q = 1$ the condition (14) is equivalent to

$$\sup_{t \neq 0} \left| \int_{-1/|t|}^{1/|t|} x\,dF(x) \right| = \sup_{a \neq 0} \left| EX I_{\{|X| \leq a\}} \right| < \infty.$$

Now we consider the case $1 < q < 2$. Suppose (14) holds. Then $J_2(t) = o(|t|^r)$ as $t \to 0$ for all $r \in (1, q)$. Therefore

$$EX = \lim_{t \to 0} \int_{-1/|t|}^{1/|t|} x\,dF(x) = \lim_{t \to 0} \frac{J_2(t)}{t} = 0.$$

Let now $EX = 0$. Then for $t \neq 0$

$$\int_{-1/|t|}^{1/|t|} x\,dF(x) = \int_{|x| \geq 1/|t|} x\,dF(x).$$

Integration by parts and the upper estimate in (7) give us

$$\left| \int_{|x| \geq 1/|t|} x\,dF(x) \right| \leq \frac{1}{|t|} P\left\{ |X| \geq \frac{1}{|t|} \right\} + \int_{|x| \geq 1/|t|} P\left\{ |X| \geq x \right\} dx.$$

Applying the upper estimate in (4) once more, we conclude that the right-hand side is $O(|t|^{q-1})$ as $t \to 0$, which implies (14). □

4. Estimates for the distributions of the sums. The result of this subsection will be used to prove Theorems 2 and 3, but it has an independent value.

Lemma 4. *Let* $\{X_k\}_{k=1}^{\infty}$ *be i.i.d.r.v.s and* $0 < q < 2$. *Suppose* (4) *holds,* $EX_1 = 0$ *if* $q > 1$ *and*

$$\sup_{a \neq 0} \left| EX_1 I_{\{|X_1| \leq a\}} \right| < \infty$$

if $q = 1$. *Then there are positive constants* u, v *and* w *such that for every* $a_k \in \mathbf{R}$, $n \in N$ *and* $x \geq w$

$$ux^{-q} \leq P\left\{ \left| \sum_{k=1}^{n} a_k X_k \right| \left(\sum_{k=1}^{n} |a_k|^q \right)^{-1/q} \geq x \right\} \leq vx^{-q}. \tag{15}$$

In addition, the upper estimate in (4) together with the other conditions of the lemma imply the upper estimate in (15).

We break the proof into several steps. Without loss of generality we may assume

$$\sum_{k=1}^{n} |a_k|^q = 1. \tag{16}$$

Let $f(t)$ be the characteristic function corresponding to the r.v. X_1. Then the sum $\sum_{k=1}^{n} a_k X_k$ has the characteristic function

$$g(t) = \prod_{k=1}^{n} f(a_k t). \tag{17}$$

We show that (7) holds for $g(t)$ with constants independent of a_k and n, which together with Lemma 2 implies (15).

Proposition 2. *Let* $\alpha_k,\ \beta_k \in \mathbf{R}$ *and* $|\alpha_k| \leq 1$ $(1 \leq k \leq n)$. *Then*

$$\left| 1 - \prod_{k=1}^{n} (\alpha_k + i\beta_k) \right| \leq \left| 1 - \prod_{k=1}^{n} \alpha_k \right| + \exp\left(\sum_{k=1}^{n} |\beta_k| \right) - 1.$$

Proof: We have

$$\left| 1 - \prod_{k=1}^{n} (\alpha_k + i\beta_k) \right| \leq \left| \prod_{k=1}^{n} \alpha_k \right| + \left| \prod_{k=1}^{n} (\alpha_k + i\beta_k) - \prod_{k=1}^{n} \alpha_k \right|.$$

Expanding the parentheses, we get

$$\prod_{k=1}^{n} (\alpha_k + i\beta_k) - \prod_{k=1}^{n} \alpha_k = i\beta_n \prod_{k=1}^{n-1} (\alpha_k + i\beta_k) + i\alpha_n \beta_{n-1} \prod_{k=1}^{n-2} (\alpha_k + i\beta_k)$$

$$+ i\alpha_n \alpha_{n-1} \beta_{n-2} \prod_{k=1}^{n-3} (\alpha_k + i\beta_k) + \cdots + i\alpha_n \alpha_{n-1} \cdots \beta_1.$$

Since $|\alpha_k| \leq 1$, it follows that

$$\left| \prod_{k=1}^{n} (\alpha_k + i\beta_k) - \prod_{k=1}^{n} \alpha_k \right| \leq \sum_{k=1}^{n-1} |\beta_{n-k+1}| \prod_{j=1}^{n-k} (1 + |\beta_j|) + |\beta_1|$$

$$= \prod_{k=1}^{n} (1 + |\beta_k|) - 1.$$

As $1 + |x| \leq \exp(|x|)$, then

$$\prod_{k=1}^{n} (1 + |\beta_k|) \leq \exp\left(\sum_{k=1}^{n} |\beta_k| \right).$$

This implies the needed inequality. $\square$

Proposition 3. *Let (7) and (12) be fulfilled. Then there are positive constants α_1, β_1 and γ_1 such that if $|t| \leq \gamma_1$, then*

$$\exp\left(-\beta_1 |t|^q\right) \leq \mathrm{Re} f(t) \leq |f(t)| \leq \exp\left(-\alpha_1 |t|^q\right). \tag{18}$$

In addition, the upper estimate in (7) implies the lower estimate in (18).

Proof: It follows from (7) and the well known inequality $1 - x < \exp(-x) < 1 - x/2$, where $0 < x < 1$, that there is $\delta > 0$ such that

$$\exp\left(-2\beta |t|^q\right) \leq \mathrm{Re} f(t) \leq \exp\left(-\alpha |t|^q\right)$$

for $|t| < \delta$, where α and β are the constants from (7). This together with (12) gives us the estimate

$$|f(t)| \leq \left(\exp\left(-2\alpha |t|^q\right) + \mu^2 |t|^{2q}\right)^{1/2},$$

where $|t| < \min\{\delta, \nu\}$.

If $h > 0$, then

$$\exp(-x) + hx^2 < 1 - \frac{x}{2} + hx^2 < 1 - \frac{x}{4} < \exp\left(-\frac{x}{4}\right)$$

for small enough positive x. Therefore there exist positive constants α_1 and δ_1 such that

$$\exp\left(-2\alpha |t|^q\right) + \mu^2 |t|^{2q} \leq \exp\left(-2\alpha_1 |t|^q\right)$$

for $|t| < \delta_1$. This and the above implies the needed inequalities. One can see that the lower estimate in (18) follows from the upper estimate in (7). $\square$

Proof of Lemma 4: According to Lemmas 2 and 3, from the conditions of Lemma 4 the estimates (7) and (12) follow. Suppose that $g(t)$ is defined by the formula (17) and (16) holds. Then $|a_k| \leq 1$ and we obtain, using (18),

$$1 - \mathrm{Re} g(t) \geq 1 - |g(t)| = 1 - \left|\prod_{k=1}^{n} f(a_k t)\right|$$

$$\geq 1 - \exp\left(-\alpha_1 \sum_{k=1}^{n} |a_k t|^q\right) = 1 - \exp\left(-\alpha_1 |t|^q\right), \tag{19}$$

where $|t| \leq \gamma_1$.

Denote $c(t) = \mathrm{Re} f(t)$ and $d(t) = \mathrm{Im} f(t)$. According to (17) and Proposition 2,

$$1 - \mathrm{Re} g(t) \leq |1 - g(t)| = \left|1 - \prod_{k=1}^{n} (c(a_k t) + i d(a_k t))\right|$$

$$\leq \left|1 - \prod_{k=1}^{n} c(a_k t)\right| + \exp\left(\sum_{k=1}^{n} |d(a_k t)|\right) - 1.$$

Since $|c(t)| \leq 1$, then from (18) and (16)

$$0 \leq 1 - \prod_{k=1}^{n} c(a_k t) \leq 1 - \prod_{k=1}^{n} \exp\left(-\beta_1 |a_k t|^q\right) = 1 - \exp\left(-\beta_1 |t|^q\right).$$

Applying (12), we get for $|t| \leq \nu$

$$\sum_{k=1}^{n} |d(a_k t)| \leq \mu \sum_{k=1}^{n} |a_k t|^q = \mu |t|^q.$$

The last estimates and (19) yield that there exist positive constants α_2, β_2 and γ_2, independent of n and a_k, such that

$$\alpha_2 |t|^q \leq 1 - Reg(t) \leq \beta_2 |t|^q \tag{20}$$

if $|t| \leq \gamma_2$. These estimates and Lemma 2 give us (15). According to Lemma 2 and Proposition 3, the upper estimate in (4) implies the same estimate in (7) and the lower bound in (18). Reasoning as above, we get the right-hand side inequality in (20). Using Lemma 2 once more, we obtain the same estimate in (15). $\square$

5. Convergence of the norms.

Lemma 5. *Let the condition (3) be fulfilled, $1 < q < 2$ and r.v.s Y_n be weakly convergent to the r.v. Y. Suppose there exists a constant $b > 0$ such that for all $x > 0$ and $n \in \mathbf{N}$*

$$P\{|Y_n| \geq x\} \leq bx^{-q}. \tag{21}$$

Then $Y, Y_n \in \mathbf{E}$ and $\|Y_n\|_{\mathbf{E}} \to \|Y\|_{\mathbf{E}}$.

Proof: The assumptions of the lemma imply that the r.v. Y satisfies (21) with the same constant. Since $Z_q \in \mathbf{E}$, Proposition 1.2 gives that Y_n and Y are contained in the space $\mathbf{E}$.

Let $a > 0$. The estimate (21) and Proposition 1.2 give us

$$\left\|Y_n I_{\{|Y_n| \geq a\}}\right\|_{\mathbf{E}} \leq B \left\|Z_q I_{\{|Z_q| \geq a\}}\right\|_{\mathbf{E}},$$

where B is independent of a and n. From here and (3)

$$\lim_{a \to \infty} \sup_n \left\|Y_n I_{\{|Y_n| \geq a\}}\right\|_{\mathbf{E}} = 0.$$

So, for fixed $\epsilon > 0$ there exists $a > 0$ such that

$$\left\|Y_n I_{\{|Y_n| \geq a\}}\right\|_{\mathbf{E}} \leq \epsilon$$

for all $n \in \mathbb{N}$ and we may conclude that for all integers n

$$\|Y_n\|_{\mathbf{E}} - \epsilon \leq \left\|Y_n I_{\{|Y_n|<a\}}\right\|_{\mathbf{E}} \leq \|Y_n\|_{\mathbf{E}}.$$

Since the r.v. Y satisfies (21), the same inequality is true for Y.

The condition (3) implies $\mathbf{E} \neq L_\infty(\Omega)$. Hence $\phi_{\mathbf{E}}(t) \to 0$ as $t \to 0$ (see [48]). Using this, it is not difficult to verify that

$$\lim_{n \to \infty} \left\|Y_n I_{\{|Y_n|<a\}}\right\|_{\mathbf{E}} = \left\|Y I_{\{|Y|<a\}}\right\|_{\mathbf{E}}.$$

From here and the previous

$$\|Y\|_{\mathbf{E}} - \epsilon \leq \varliminf_{n \to \infty} \|Y_n\|_{\mathbf{E}} \leq \varlimsup_{n \to \infty} \|Y_n\|_{\mathbf{E}} \leq \|Y\|_{\mathbf{E}} + \epsilon.$$

Putting $\epsilon \to 0$, we obtain the needed relation. $\square$

6. Some more inequalities for characteristic functions. Here we show that (1) implies (7).

Lemma 6. *Let $\{X_k\}_{k=1}^{\infty}$ be i.i.d.r.v.s, $f(t)$ be the corresponding characteristic function and let $0 < p < q < 2$ be fixed. Suppose*

$$C_{p,q} \equiv \sup_n \left| n^{1/q} \sum_{k=1}^{n} X_k \right|^p < \infty. \tag{22}$$

Then (12) and the upper estimate in (7) are true.

First we prove some auxiliary results. Since $f(t)$ is continuous and $f(0) = 1$, there exists $\log f(t)$ near zero. We consider the branch of the logarithm for which $\log(1) = 0$ and put

$$\phi(t) = -\frac{\log f(t)}{|t|^q}.$$

Therefore, there is $\delta > 0$ such that

$$f(t) = \exp\left((-|t|^q \, \phi(t))\right) \tag{23}$$

for $|t| \leq \delta, t \neq 0$. To prove Lemma 6 we have to set bounds for $\phi(t)$ near zero.

Let's denote

$$S_n = n^{-1/q} \sum_{k=1}^{n} X_k. \tag{24}$$

This sum has the characteristic function

$$f_n(t) = \left(f(tn^{-1/q}) \right)^n . \tag{25}$$

The condition (22) yields that the sequence $\{S_n\}_{n=1}^\infty$ is weakly compact (see [35]). Let's denote the collection of all characteristic functions related to the limit distributions of $\{S_n\}_{n=1}^\infty$ by Λ. Put

$$\Psi = \Lambda \bigcup \{f_n\}_{n=1}^\infty .$$

By Y_h we denote a r.v. with the characteristic function $h(t)$. If $h \in \Psi$, then (22) and the well-known properties of absolute moments (see [35]) imply that $E|Y_h| \le C_{p,q}$. Therefore the collection $\{Y_h : h \in \Psi\}$ is also weakly compact.

Proposition 4. *Let (22) be fulfilled. Then there is $\epsilon > 0$ for which*

$$\nu(\epsilon) = \inf \{\operatorname{Re}h(t) : |t| \le \epsilon, h \in \Psi\} > 1/2.$$

Proof: Suppose the contrary holds. Then there are $t_k \searrow 0$ and $h_k \in \Psi$ such that $\operatorname{Re}h_k(t_k) \le 1/2$ $(k \in \mathbf{N})$. By virtue of the compactness we may choose integers $k(n) \nearrow \infty$ with the property $h_{k(n)}(t) \to g(t)$ for every $t \in \mathbf{R}$, where $g(t)$ is a characteristic function. From here (see [35])

$$\lim_{n \to \infty} h_{k(n)}(t_{k(n)}) = g(0) = 1.$$

This relation contradicts the above. $\square$

Choosing $\epsilon > 0$ under the condition $\nu(\epsilon) > 1/2$ we have $\operatorname{Re}h(t) > 1/2$ for all $h \in \Psi$ and $t \in (-\epsilon, \epsilon)$. So, for every $h \in \Psi$ the function

$$\phi_h(t) = -\frac{\log(h(t))}{|t|^q}$$

is defined on $(-\epsilon, \epsilon) \backslash \{0\}$ and on this set the formula

$$h(t) = \exp\left(- |t|^q \phi_h(t)\right) \tag{26}$$

takes place.

Proposition 5. *Let $h \in \Psi$ and $f_{k(n)}(t) \to h(t)$ for each $t \in \mathbf{R}$. Then $\phi(tk(n)^{-1/q}) \to \phi_h(t)$ uniformly on each set $\{t : 0 < \alpha \le |t| \le \beta \le \epsilon\}$, where $\phi(t)$ is the function from (23).*

Proof: It follows from (23) and (25) that

$$f_n(t) = \exp\left(- |t|^q \phi\left(n^{-1/q}t\right)\right) \tag{27}$$

if $0 < |t| \leq \epsilon$. From here

$$\left| \phi\left(n^{-1/q}t\right) - \phi\left(m^{-1/q}t\right) \right| = \frac{|\log f_n(t) - \log f_m(t)|}{|t|^q} \leq \frac{|f_n(t) - f_m(t)|}{|t|^q |d_{m,n}(t)|} \,,$$

where $d_{m,n}(t)$ is a point contained in the linear segment joining the points $f_m(t)$ and $f_n(t)$. According to Proposition 2, $|f_n(t)| \geq 1/2$ if $|t| \leq \epsilon$ $(n \in \mathbf{N})$. Hence $|d_{m,n}(t)| \geq 2$.

It is well known that the convergence of characteristic functions is uniform on each finite segment (see [35]). Therefore the last estimates yield the needed assertion. $\square$

Proof of Lemma 6: According to (23), the function $\phi(t)$ is defined on $(-\epsilon, \epsilon)\backslash\{0\}$. Proposition 5 and Arzela–Ascoli's theorem (see [16]) give us

$$C \equiv \sup\left\{ \left|\phi\left(tn^{-1/q}\right)\right| : \epsilon 2^{-1/q} \leq |t| \leq \epsilon,\, n \in \mathbf{N} \right\} < \infty.$$

We show that $|\phi(t)| \leq C$ for every $t \in (-\epsilon, \epsilon)\backslash\{0\}$. It is not difficult to verify that for such t there is an integer n with the property $\epsilon 2^{-1/q} \leq \left|tn^{1/q}\right| \leq \epsilon$. From this and the above

$$|\phi(t)| = \left| \phi\left(n^{-1/q}\left(tn^{1/q}\right)\right) \right| \leq C$$

This bound and (23) give us $1 - f(t) = O\left(|t|^q\right)$ as $t \to 0$, which implies Lemma 6. $\square$

Lemma 7. *Let one of the following conditions be fulfilled:*
 (a) $1 < q < 2$ and (3) holds;
 (b) $0 < p < q < 2$ and $\mathbf{E} = L_p(\Omega)$.
Let $\{X_k\}_{k=1}^{\infty} \subset \mathbf{E}$ be a sequence of i.i.d.r.v.s such that (1) holds and, if $\mathbf{E} = L_p(\Omega)$ and $p < q = 1$,

$$\inf_n \left\| \sum_{k=1}^{n} (X_{2k} - X_{2k-1}) \right\|_{\mathbf{E}} > 0. \tag{28}$$

Then for the characteristic function $f(t)$ corresponding to X_1 the estimates (7) and (12) are true.

If follows from (1) and Proposition 1.1 that in the case (a) the condition (22) holds, where $1 = p < q < 2$. This condition follows from (1) in the case (b). According to Lemma 6, the upper estimates in (7) and (12) are true. Hence, we have only to prove the lower estimate in (7).

Proposition 6. *Suppose $h_m \in \Psi$ and Y_{h_m} are weakly convergent to Y . Then $Y \in \mathbf{E}$ and*

$$\lim_{m \to \infty} \|Y_{h_m}\|_{\mathbf{E}} = \|Y\|_{\mathbf{E}} .$$

Proof: The upper estimates in (7) and (12) give us the conditions of Lemma 4 (see Lemmas 2 and 3). So, for the sum (24) the inequality (21) holds with a constant independent of n. According to the definition of the collection Ψ, the same estimate is true for every r.v. Y_h, $h \in \Psi$.

Suppose (a) holds. Then the needed assertion follows directly from Lemma 5. Let (b) be fulfilled. According to (21)

$$C_r \equiv \sup_{h \in \Psi} E |Y_h|^r < \infty \tag{29}$$

for every $r \in (p, q)$. From here and the theorem on the convergence of moments (see [35]) the needed relation follows. $\square$

Proposition 7. *Let $h \in \Psi$ and $\{Y_k\}_{k=1}^{\infty}$ be independent r.v.s equidistributed with Y_h . Then these r.v.s satisfy the condition of Lemma 7.*

Proof: For each $h \in \Psi$ there are integers $n(j) \nearrow \infty$ such that $S_{n(j)}$ weakly converges to Y_h, where S_n is determined by the formula (24). Let's fix an integer m and consider

$$U_m = m^{-1/q} \sum_{k=1}^{m} Y_k . \tag{30}$$

We have

$$S_{mn(j)} = m^{-1/q} \sum_{k=1}^{m} n(j)^{-1/q} \sum_{i=1}^{n(j)} X_{(k-1)n(j)+i} .$$

Hence $S_{mn(j)} \to U_m$ weakly as $j \to \infty$, and according to Proposition 6 $\|S_{mn(j)}\|_{\mathbf{E}} \to \|U_m\|_{\mathbf{E}}$. From (1)

$$C_1 \leq \|S_{mn(j)}\|_{\mathbf{E}} \leq C_2$$

for all j. So, the same estimate holds for U_m, which is equivalent to (1) for Y_k.

In the case $p < q = 1$ it is similarly proved that (28) holds for the sequence in question. $\square$

As usual we call a r.v. X and the corresponding characteristic function degenerate if $P\{x = a\} = 1$ for some $a \in \mathbf{R}$.

Proposition 8. *Suppose that for some $0 < \mu < \nu$ and for the functions f_n determined by (25)*

$$\sup \{|f_n(t)| : \mu \le t \le \nu , n \in \mathbf{N}\} = 1.$$

Then there are $h \in \Psi$ and $t_0 \in [\mu, \nu]$ such that $|h(t_0)| = 1$ and h is non-degenerate.

Proof: There are $t_k \in [\mu, \nu]$ and integers $n(k) \nearrow \infty$ such that $|f_{n(k)}(t_k)| \to 1$. As mentioned above, (29) yields the weak compactness of $\{Y_h : h \in \Psi\}$. Passing to a subsequence, we may assume that $t_k \to t_0 \in [\mu, \nu]$ and $f_{n(k)}(t) \to h(t)$ for all $y \in \mathbf{R}$ and some $h \in \Psi$. From here (see [35])

$$|h(t_0)| = \lim_{k \to \infty} |f_{n(k)}(t_k)| = 1.$$

Suppose Y_h is degenerate and consider independent r.v.s $\{Y_k\}_{k=1}^{\infty}$ equidistributed with Y_h. Then Y_k equals a constant almost surely. But it is not difficult to verify that then (1) does not hold if $q \ne 1$, and (28) is not true if $q = 1$. This contradicts Proposition 7. $\square$

Let's denote

$$J_m = \left[(2m)^{-1/q} , m^{-1/q}\right] \tag{31}$$

Proposition 9. *There is an integer m such that*

$$\sup \{|f_n(t)| : t \in J_m , n \in \mathbf{N}\} = \gamma < 1 .$$

Proof: Suppose the contrary holds. Then, according to Proposition 8, for each m there are non-degenerate $h_m \in \Psi$ and $t_m \in J_m$ such that

$$|h_m(t_m)| = 1.$$

Hence h_m corresponds to a lattice distribution. If a_m is the maximal step of this distribution, then (see [43], Ch.1) $t_m \ge 2\pi/a_m$ and $a_m \ge 2\pi/t_m \ge 2\pi m^{1/q}$. As the collection $\{Y_h : h \in \Psi\}$ is weakly compact, there are integers $m(k) \nearrow \infty$ for which $h_{m(k)}(t) \to h(t) \in \Psi$ for every real t. It is not difficult to prove that $a_m \to \infty$ implies $|h(t)| \equiv 1$. So, the r.v. Y_h is degenerate. As in the proof of Proposition 8, a contradiction follows. $\square$

The next assertion is easily verified.

Proposition 10. *For every $0 < \mu < \nu$ and integer j there is $\epsilon > 0$ such that*

$$(0,\epsilon) \subset \bigcup_{n \ge j} \left[\mu n^{-1/q} , \nu n^{-1/q}\right] .$$

Proof of Lemma 7: Let's denote $u(t) = \text{R}ef(t)$. According to (25) and Proposition 9, for some integer m and every $t \in J_m$

$$\left| u\left(tn^{-1/q} \right) \right|^n \leq \left| f\left(tn^{-1/q} \right) \right|^n = |f_n(t)| \leq \gamma < 1.$$

Since $u(0) = 1$, there is $\delta > 0$ such that $u(x) > 0$ if $|x| < \delta$. In addition, for some integer j the inequality $tn^{-1/q} < \delta$ holds if $n \geq j$ and $t \in J_m$, which yields $n \log \left(u \left(tn^{-1/q} \right) \right) \leq \log(\gamma) < 0$ for such n and t. From here and (31)

$$- \left(tn^{-1/q} \right)^{-q} \log \left(u \left(tn^{-1/q} \right) \right) \geq -t^{-q} \log(\gamma) \geq -m \log(\gamma) = \alpha > 0 \,.$$

Therefore $u \left(tn^{-1/q} \right) \leq \exp \left(-\alpha t^q n^{-1} \right)$.

Proposition 10 implies that the union of the segments J_{mn} $(n \geq j)$ contains some interval $(0, \epsilon)$. If $t \in (0, \epsilon)$, then $tn^{1/q} \in J_m$ for some $n \geq j$. Taking into account the previous, we get

$$u(t) = u\left(n^{-1/q} \left(tn^{1/q} \right) \right) \leq \exp \left(\frac{-\alpha \left(tn^{1/q} \right)^q}{n} \right) = \exp \left(-\alpha t^q \right).$$

It follows from (9) that $u(t)$ is even. Hence the last inequality yields the lower estimate in (7). $\square$

7. Proof of Theorems 2 and 3. Let the conditions of Theorem 2 be fulfilled. According to Lemma 7, the estimates (7) and (12) are true. From Lemmas 2 and 3, $EX_1 = 0$ and (4) holds.

If the right-hand side of (1) holds, then Proposition 1.1 implies (22), where $p = 1 < q < 2$. Using Lemmas 2 and 3 we get the second part of Theorem 2. Theorem 3 follows immediately from Lemma 4. $\square$

8. Proof of Theorem 4. Let $\{Y_k\}_{k=1}^{\infty}$ be symmetric i.i.d.r.v.s with the q-stable distribution. It follows from (1.18) that Y_1 satisfies the estimate (4) for some positive constants a, b and c. We choose numbers α_j, β_j and x_j from (c, ∞) under the conditions

$$x_j \geq 4j^{1-1/q}\alpha_j, \tag{32}$$

$$\beta_j \geq \left(\frac{2bj}{a} \right)^{1/q} x_j, \tag{33}$$

$$\lim_{j \to \infty} \frac{\beta_j}{\alpha_{j+1}} = 0. \tag{34}$$

Let's put

$$X_k = \sum_{j=1}^{\infty} Y_k I_{\{\alpha_j \leq |Y_k| < \beta_j\}}. \tag{35}$$

It is clear that $\{X_k\}_{k=1}^{\infty}$ are symmetric i.i.d.r.v.s contained in the space $L_{q,\infty}(\Omega)$.

From (34) and (35) $P\{|X_k| \geq \beta_j\} = P\{|X_k| \geq \alpha_{j+1}\}$. The formula (35) gives us the estimate $P\{|X_k| \geq x\} \leq P\{|Y_k| \geq x\}$, which holds for every $x > 0$. From here and (4)

$$P\{|X_k| \geq \beta_j\} \leq P\{|Y_k| \geq \alpha_{j+1}\} \leq b\alpha_{j+1}^{-q}$$

Taking into account (34), we obtain

$$\beta_j^q P\{|X_k| \geq \beta_j\} \leq b\left(\frac{\beta_j}{\alpha_{j+1}}\right)^q \to 0$$

as $j \to \infty$. Hence X_k does not satisfy the lower estimate in (4).

Since X_k satisfies the upper estimate in (4), then according to Theorem 3, the upper estimate in (5) takes place, where $\mathbf{E} = L_{p,\infty}(\Omega)$. We prove the lower estimate.

Let's put for fixed n and $k \in \mathbf{N}$

$$Z_{k,n} = (Y_k - X_k) I_{\{|Y_k| - X_k\} \leq \alpha_n}, \quad U_{k,n} = Y_k - X_k - Z_{k,n}.$$

Let a collection of real numbers $\{a_k\}_{k=1}^{n}$ satisfy (16) and denote by S_j ($j = 1, 2, 3$) the linear combinations with the coefficients $\{a_k\}_{k=1}^{n}$ of the r.v.s $\{X_k\}$, $\{Z_{k,n}\}$ and $\{U_{k,n}\}$ respectively. Then

$$\sum_{k=1}^{n} a_k Y_k = S_1 + S_2 + S_3.$$

From (1.20) and (16) $S_1 + S_2 + S_3 \overset{d}{=} Y_1$. This and the lower estimate in (4) for the r.v. Y_1 imply that

$$ax^{-q} \leq P\{|Y_1| \geq x\} \leq \sum_{j=1}^{3} P\left\{|S_j| \geq \frac{x}{3}\right\} \tag{36}$$

for $x > c$. Using (16), (32) and Hölder's inequality, we get

$$\sum_{k=1}^{n} |a_k| \leq n^{1-1/q}\left(\sum_{k=1}^{n} |a_k|^q\right)^{1/q} = n^{1-1/q} < \frac{x_n}{4\alpha_n}.$$

It follows from the definition that $|Z_{k,n}| \leq \alpha_n$, which yields $|S_2| \leq x_n/4$ and $P\{|S_2| \geq x_n/3\} = 0$. From (35), (33) and (4) we obtain

$$P\{U_{k,n} \neq 0\} = P\{|Y_k - X_k| \geq \alpha_n\} \leq P\{|Y_k| \geq \beta_n\} \leq b\beta_n^{-q} \leq \frac{ax_n^{-q}}{2n}.$$

Hence

$$P\left\{|S_3| \geq \frac{x_n}{3}\right\} \leq \sum_{k=1}^{n} P\left\{U_{k,n} \neq 0\right\} \leq \frac{a x_n^{-q}}{2}.$$

These results for S_1 and S_2 and the estimate (36) yield $P\left\{|S_1| \geq x_n/3\right\} \geq a x_n^{-q}/2$. From here and (1.2), under the condition (16),

$$\left\|\sum_{k=1}^{n} a_k X_k\right\|_{p,\infty}^{*} \geq \frac{1}{3} \frac{a}{2^{1/q}}$$

and we get the lower estimate in (5). $\square$

9. The estimates for the exponents. Here we show that the conditions of Theorem 5 imply $0 < p < q < 2$. The following result is proved in [47].

Lemma 8. *Let $\{X_k\}_{k=1}^{\infty}$ be i.i.d.r.v.s, $0 < p < 2$ and $E\,|X_1|^p < \infty$ Suppose $EX_1 = 0$ if $p \geq 1$. Then*

$$E\left|\sum_{k=1}^{n} X_k\right|^p = o(n) \tag{37}$$

as $n \to \infty$.

Proof: First let $1 \leq p < 2$ and r.v.s X_k be symmetric. Put for $a > 0$

$$X_{k,a} = X_k I_{\{|X_k| < a\}} \quad , \quad Z_{k,a} = X_k - X_{k,a}. \tag{38}$$

These r.v.s are symmetric. Since $p < 2$, then

$$E\left|\sum_{k=1}^{n} X_{k,a}\right|^p \leq \left(E\left(\sum_{k=1}^{n} X_{k,a}\right)^2\right)^{p/2} = \left(\sum_{k=1}^{n} E\left(X_{k,a}^2\right)\right)^{p/2} \leq (an)^p.$$

From here

$$E\left|n^{-1/p}\sum_{k=1}^{n} X_{k,a}\right|^p \leq a^p n^{p/2-1} \to 0.$$

According to von Bahr–Esseen's inequality [2],

$$E\left|\sum_{k=1}^{n} Z_{k,a}\right|^p \leq 2\sum_{k=1}^{n} E\,|Z_{k,a}|^p = 2nE\,|Z_{1,a}|^p.$$

Using the well-known inequality

$$E\,|U + V|^p \leq b(p)\left(E\,|U|^p + E\,|V|^p\right), \tag{39}$$

where $b(p) = \max\left\{1, 2^{p-1}\right\}$, we obtain

$$\varlimsup_{n \to \infty} E\left|n^{-1/p}\sum_{k=1}^{n} X_k\right|^p \le 2b(p)E\,|Z_{1,a}|^p\,.$$

If $a \to \infty$, then $E\,|Z_{1,a}|^p \to 0$ and we get (37).

Now we eliminate the assumption of symmetry. Let's put $Y_k = X_{2k} - X_{2k-1}$. Since $EX_1 = 0$ then from Proposition 1.11

$$E\left|\sum_{k=1}^{n} X_k\right|^p \le E\left|\sum_{k=1}^{n} Y_k\right|^p,$$

which yields (37).

In the the case $0 < p < 1$ we have

$$E\left|\sum_{k=1}^{n} X_{k,a}\right|^p \le E\left(\sum_{k=1}^{n}|X_{k,a}|\right)^p \le (an)^p.$$

As $p < 1$, then

$$E\left|n^{-1/p}\sum_{k=1}^{n} X_{k,a}\right| \le a^p n^{p-1} \to 0.$$

From (39)

$$E\left|\sum_{k=1}^{n} Z_{k,a}\right|^p \le \sum_{k=1}^{n} E\,|Z_{k,a}|^p = nE\,|Z_{1,a}|^p$$

and (37) follows. $\square$

Lemma 9. *If the conditions of Theorem 5 hold, then $0 < p < q < 2$.*

Proof: As in the proof of Lemma 1, we get that the estimate (6) holds for the spaces $L_p(\Omega)$ ($p < 1$). From here and (1) $q \le 2$. Since $q \ne 2$, then we have $q < 2$.

Now we show that $p < 2$. Suppose $p \ge 2$. Applying Theorem 1, we get $q \ge 1$. Let $q > 1$. Then by the same theorem $EX_1 = 0$. The lower estimate in (1) and Rosenthal's inequality give us

$$C_1 n^{1/q} \le \left\|\sum_{k=1}^{n} X_k\right\|_p \le C(p)\max\left\{\left(\sum_{k=1}^{n}\|X_k\|_p^p\right)^{1/p}, \left(\sum_{k=1}^{n}\|X_k\|_2^2\right)^{1/2}\right\}$$

$$= C(p)\max\left\{n^{1/p}\,\|X_1\|_p\,, n^{1/2}\,\|X_1\|_2\right\},$$

which contradicts the strong inequality $q < 2$.

In the case $q = 1$ we have to apply Rosenthal's inequality to the r.v.s $\{X_{2k} - X_{2k-1}\}_{k=1}^{\infty}$, which together with (28) leads to a contradiction.

So, $p < 2$. Now we prove that $p \leq q$. Let, at first, $q \neq 1$. If $p \geq 1$, then from Theorem 1, $q > 1$ and $EX_1 = 0$. If $p < 1$ then $b(p) = 1$ in (39). Using von Bahr–Esseen's inequality for $p \geq 1$ and (39) for $p < 1$ and applying (1), we get

$$C_1 n^{p/q} \leq E \left| \sum_{k=1}^{n} X_k \right|^p \leq 2 \sum_{k=1}^{n} E |X_k|^p = 2nE |X_1|^p .$$

From here $p \leq q$.

Let $q = 1$. Considering the r.v.s $\{X_{2k} - X_{2k-1}\}_{k=1}^{\infty}$, we obtain as above, $p \leq q = 1$.

Lastly we show that $p \neq q$. If $p = q > 1$, then $EX_1 = 0$, which follows from Theorem 1 . Using Lemma 8, we get a contradiction to the left hand-side of (1). In the case $p = q = 1$ we apply Lemma 8 to the r.v.s $\{X_{2k} - X_{2k-1}\}_{k=1}^{\infty}$ and obtain a contradiction to (28). Lemma 8 shows that in the case $p = q < 1$ the lower estimate in (1) is not true. $\square$

10. Proof of Theorems 5 and 6. Let the conditions of Theorem 5 be fulfilled. Then Lemma 9 yields $0 < p < q < 2$. Lemma 7 gives us the estimates (7) and (12). Applying Lemmas 2 and 3, we get the assertions 2)–4).

Theorem 6 follows directly from Lemma 4. $\square$

2. l_2-estimates

1. Results. In this section we consider estimates of the types (1) and (5) for $q = 2$. From Lemma 1, in this case the left-hand side of (5) holds. Hence, we have to study the right-hand side of (5) only, i.e. the estimate

$$\left\| \sum_{k=1}^{n} a_k X_k \right\|_{\mathbf{E}} \leq D \left(\sum_{k=1}^{n} a^2 \right)^{1/2} , \tag{40}$$

where D is a constant independent of n and a_k.

Theorem 1 and (40) yield $EX_1 = 0$. Lemma 2.6 and Proposition 1.1 give us $EX_1^2 < \infty$. Putting $a_k = n^{-1/2}$ ($1 \leq k \leq n$), we obtain

$$\sup_{n} \left\| n^{-1/2} \sum_{k=1}^{n} X_k \right\|_{\mathbf{E}} \leq D .$$

It follows from the Central Limit Theorem that the normed sums $S_n = n^{-1/2} \sum_{k=1}^{n} X_k$ are weakly convergent to a r.v. with the normal distribution.

Using Proposition 1.5 we conclude that the space $\mathbf{E}''$ contains a r.v. with such a distribution.

So, the necessary conditions for (40) to be fulfilled are the following:

1) the space $\mathbf{E}''$ contains a r.v. with the normal distribution;
2) $EX_1^2 < \infty$;
3) $EX_1 = 0$.

The question of the sufficiency of these conditions is open. But under some additional assumptions the answer is positive.

Let $L_2(\Omega) \subset \mathbf{E}$. Then $\|X\|_{\mathbf{E}} \leq C\|X\|_2$ for every $X \in L_2(\Omega)$ and some constant C, which together with 2) and 3) yields (40). If $\mathbf{E}$ has von Bahr—Esseen's 2-property, then (40) also holds. Applying Theorem 2.7, we get the estimate (40) for the spaces $\mathbf{E} = L_{p,q}(\Omega)$, where $p > 2$ and $q \geq 2$. It was proved in [15] that (40) is true in the case $p > 2$ and $q < 2$. The results of the section 2.4 permit us to conclude that the considered estimate holds for the Orlicz spaces $L_{N_p}(\Omega)$, where $0 < p \leq 2$.

We introduce a new notation. For $\mathbf{a} = \{a_k\}_{k=1}^{\infty} \in l_2$ and a r.v. X we put

$$Q_{\mathbf{a}}(x) = \sum_{k=1}^{\infty} P\{|a_k X| \geq x\}. \tag{41}$$

Suppose $X \in L_2(\Omega)$. Then $P\{|X| \geq x\} \leq Cx^{-2}$, where C is a constant. So,

$$Q_{\mathbf{a}}(x) \leq C \left(\sum_{k=1}^{\infty} a_k^2\right) x^{-2} < \infty$$

for all $x > 0$.

Definition 1. *We say that a r.v.* X *has the* $A_2(\mathbf{E})$-*property* $(X \in A_2(\mathbf{E}))$ *if for all* $\mathbf{a} \in l_2$ *the r.i. space* $\mathbf{E}$ *contains all r.v.s* Y *such that* $P\{|Y| \geq x\} \leq CQ_{\mathbf{a}}(x)$ *for every* $x > 0$ *and a constant* C.

It is obvious that $X \in A_2(\mathbf{E})$ implies $X \in \mathbf{E}$.

We describe some conditions which yield the $A_2(\mathbf{E})$-property. Suppose that there are positive constants b and c such that

$$P\{|X| \geq xy\} \leq cy^{-2}P\{|X| \geq x\} \tag{42}$$

for $x > b$ and all positive y. From (42)

$$Q_{\mathbf{a}}(x) \leq c \left(\sum_{k=1}^{\infty} a_k^2\right) P\{|X| \geq x\}.$$

This and Proposition 1.2 imply that if $X \in \mathbf{E}$, then $X \in A_2(\mathbf{E})$.

Let $P\{|X| \geq x\} = \phi(x)x^{-2}$ and $\phi(x)$ be decreasing for large enough x. Then (42) holds. It is easy to show that if $P\{|X| \geq x\}$ is a function with regular variation at infinity and the exponent $p > 2$ (see [20]), then the estimate (42) is also true.

If $\mathbf{E}$ satisfies the upper 2-estimate, then $X \in \mathbf{E}$ implies that $X \in A_2(\mathbf{E})$.

The main result of this section is the following.

Theorem 7. *Let* **E** *have the Kruglov property and* $\{X_k\}_{k=1}^{\infty}$ *be i.i.d.r.v.s, such that* $X_1 \in A_2(\mathbf{E})$, $EX_1^2 < \infty$ *and* $EX_1 = 0$. *Then (40) holds.*

2. On infinite divisible distributions. The proof of Theorem 7 is based on the next result which has an independent interest.

We recall that a r.v. X has an infinite divisible distribution if for every integer n there are i.i.d. r.v.s $\{X_{k,n}\}_{k=1}^{n}$ such that

$$X \stackrel{d}{=} \sum_{k=1}^{n} X_{k,n} \ .$$

The corresponding characteristic function is determined by Lévy-Khinchin's formula

$$f(t) = \exp\left(i\gamma t + \int_{-\infty}^{\infty} \left(e^{itx} - 1 - \frac{itx}{1+x^2} \right) \frac{1+x^2}{x^2} dG(x) \right), \qquad (43)$$

where γ is a real constant and $G(x)$ is a bounded non-decreasing function on **R**, which is called the Lévy-Khinchin spectral function. We may assume $G(-\infty) = 0$.

Let's put

$$F_G(x) = \frac{G(x)}{G(\infty)} \ . \qquad (44)$$

This is a distribution function. We recall $X \in \mathcal{L}(F)$ denotes that the r.v. X has the distribution F.

Lemma 10. *Suppose the r.v.* X *has an infinite divisible distribution,* $G(x)$ *is the corresponding Lévy-Khinchin function and* $Y \in \mathcal{L}(F_G)$. *Let a r.i. space* **E** *have the Kruglov property. Then the conditions* $X \in \mathbf{E}$ *and* $Y \in \mathbf{E}$ *are equivalent.*

Proof. The implication $(Y \in \mathbf{E}) \Rightarrow (X \in \mathbf{E})$: Let's put

$$\Psi(t, x) = \left(e^{itx} - 1 - \frac{itx}{1+x^2} \right) \frac{1+x^2}{x^2}, \qquad (45)$$

$$f_1(t) = \exp\left(i\gamma t + \int_{|x|<1} \Psi(x, t) dG(x) \right), \qquad (46)$$

$$f_2(t) = \exp\left(\int_{|x|\geq 1} \Psi(x, t) dG(x) \right) \qquad (47)$$

and denote by X_1 and X_2 independent r.v.s with the characteristic functions $f_1(t)$ and $f_2(t)$ respectively . Since $f(t) = f_1(t) f_2(t)$ then

$$X \stackrel{d}{=} X_1 + X_2. \qquad (48)$$

We show that X_1 and X_2 belong to $\mathbf{E}$.

The following statement is proved in [**31**]. Let V be a r.v. with an infinite divisible distribution. Suppose that the related spectral function is a constant on each of the intervals $(-\infty, -a)$ and (a, ∞), where a is a positive constant. Then

$$\lim_{x \to \infty} \frac{-\log P\{|V| \geq x\}}{x \log(1 + x)} = \frac{1}{a}.$$

Applying this result to the r.v.s X_1 and Z with the Poisson distribution and the parameter $\lambda = 1$, we get $P\{|X_1| \geq x\} \leq CP\{|Z| \geq x\}$ for all $x > 0$, where C is a constant. As mentioned above, if $\mathbf{E} \in \mathbf{K}$, then $Z \in \mathbf{E}$, which yields $X_1 \in \mathbf{E}$.

Now we show that $X_2 \in \mathbf{E}$. Let's denote

$$b = \int_{|x| \geq 1} x^{-1} dG(x)$$

and define the function $H(x)$ by the equalities

$$H(x) = \begin{cases} \int_{-\infty}^x \frac{1+y^2}{y^2} dG(y) & \text{if } -\infty < x \leq -1, \\ H(-1) & \text{if } -1 < x \leq 1, \\ H(-1) + \int_1^x \frac{1+y^2}{y^2} dG(y) & \text{if } x > 1. \end{cases}$$

Putting

$$g(t) = \exp\left(\int_{-\infty}^{\infty} \left(e^{itx} - 1\right) dH(x)\right), \tag{49}$$

we have from (47) $f_2(t) = g(t) \exp(ibt)$. Let U be a r.v. with the characteristic function $g(t)$. We have $X_2 \overset{d}{=} U + b$ and show $U \in \mathbf{E}$.

If $H(\infty) = 0$, then $U \equiv 0$. So, we may assume $H(\infty) > 0$, which implies $G(\infty) > 0$.

Let $F_H(x)$ be defined by the formula (44), where G will be replaced by H, and $Z \in \mathcal{L}(F_H)$. First we prove that the conditions $Y \in \mathbf{E}$ and $Z \in \mathbf{E}$ are equivalent, which follows from Proposition 1.2 and the next assertion.

Proposition 11. For $x \geq 1$

$$\frac{G(\infty)}{H(\infty)} P\{|Y| \geq x\} \leq P\{|Z| \geq x\} \leq \frac{2G(\infty)}{H(\infty)} P\{|Y| \geq x\} \tag{50}$$

Proof: Since $1 < (1 + y^2)y^{-2} \leq 2$ if $y \geq 1$, then for $x \geq 1$

$$P\{|Z| \geq x\} = \frac{1}{H(\infty)} \int_{|y| \geq x} \frac{1+y^2}{y^2} dG(x)$$

$$\leq \frac{2}{H(\infty)} \int_{|y| \geq x} dG(x) = \frac{2G(\infty)}{H(\infty)} P\{|Y| \geq x\}.$$

In addition,

$$P\{|Z| \geq x\} \geq \frac{1}{H(\infty)} \int_{|y| \geq x} dG(x) = \frac{G(\infty)}{H(\infty)} P\{|Y| \geq x\} \; \square$$

Now we reduce the proof to the case $H(\infty) = 1$. Let us put for $v > 0$

$$g_v(t) = \exp\left(v \int_{-\infty}^{\infty} \left(e^{itx} - 1\right) dH(x)\right)$$

and denote by Z_v a r.v. with this characteristic function.

Proposition 12. *For every r.i. space* **E** *and* $v > 0$ *the conditions* $Z \in \mathbf{E}$ *and* $Z_v \in \mathbf{E}$ *are equivalent.*

Proof: It is enough to show that $Z \in \mathbf{E}$ implies $Z_v \in \mathbf{E}$. We choose an integer n so that $n < v \leq n+1$ and suppose that $Z \in \mathbf{E}$. Let $\{Y_k\}_{k=1}^{n+1}$ be independent r.v.s equidistributed with Z, X be a r.v. with the characteristic function $g_{n+1-v}(t)$ and X and Z be independent. Then

$$X + Z_v \overset{d}{=} \sum_{k=1}^{n+1} Y_k \in \mathbf{E}.$$

From here and Proposition 1.10, $Z_v \in \mathbf{E}$. $\square$

We continue to prove Lemma 10. It follows from $Y \in \mathbf{E}$ that **E** contains a r.v. Z with the distribution function $F_H(x)$. According to Proposition 12, we may assume $H(\infty) = 1$. But at the same time $H(x) = F_H(x)$, which follows from (44). Hence, the characteristic function (49) corresponds to the distribution $\Pi(F_H)$. Since $\mathbf{E} \in \mathbf{K}$, then **E** contains a r.v. $U \in \mathcal{L}(\Pi(F_H))$. From here and the above $X_2 \in \mathbf{E}$. So, $X \in \mathbf{E}$.

The implication $(X \in \mathbf{E}) \Rightarrow (Y \in \mathbf{E})$: From Proposition 1.10 and the relation (48) $X_2 \in \mathbf{E}$. As shown above, $X_2 \overset{d}{=} U + b$, where U has the characteristic function (49). Hence $U \in \mathbf{E}$. Proposition 12 permits us to assume $H(\infty) = 1$. From here and the Kruglov property $Z \in \mathbf{E}$, where Z is a r.v. with the distribution function $H(x)$. Taking into account Proposition 11, we get $Y \in \mathbf{E}$. $\square$

3. Proof of Theorem 7. It follows from Kolmogorov's theorem about three series (see [35]) that under the conditions of Theorem 7 the series $\sum_{k=1}^{\infty} a_k X_k$ converges almost surely if $\{a_k\}_{k=1}^{\infty} \in l_2$.

The next statement is well known and has a standard proof.

Proposition 13. *The estimate (40) is fulfilled if and only if when*

$$\sum_{k=1}^{\infty} a_k X_k \in \mathbf{E}$$

for each $\{a_k\}_{k=1}^{\infty} \in l_2$.

Suppose at first that the r.v.s X_k are symmetric. Let F be the corresponding distribution and $\{Y_k\}_{k=1}^{\infty}$ be independent r.v.s with the common distribution $\Pi(F)$. It follows from (1.6) that Y_k are also symmetric. Kruglov's Theorem (see the section 1.6) yields that $EY_k^2 < \infty$. Hence, the series $\sum_{k=1}^{\infty} a_k X_k \in \mathbf{E}$ converges almost surely for every sequence $\{a_k\}_{k=1}^{\infty} \in l_2$.

Prokhorov's inequality (see the section 1.4) yields that for all $x > 0$, $a_k \in \mathbf{R}$ and integers n

$$P\left\{\left|\sum_{k=1}^{\infty} a_k X_k\right| \geq x\right\} \leq 8P\left\{\left|\sum_{k=1}^{\infty} a_k Y_k\right| \geq \frac{x}{2}\right\}.$$

So, the condition $\sum_{k=1}^{\infty} a_k Y_k \in \mathbf{E}$ implies $\sum_{k=1}^{\infty} a_k X_k \in \mathbf{E}$. Taking into account Proposition 13, we conclude that the problem is reduced to the proof that $\sum_{k=1}^{\infty} a_k Y_k \in \mathbf{E}$ for every sequence $\{a_k\}_{k=1}^{\infty} \in l_2$.

The outline of the proof is the following. The distribution of the r.v.s Y_k is infinite divisible and, therefore, the distribution of the sum

$$S = \sum_{k=1}^{\infty} a_k Y_k \tag{51}$$

is the same. Denote the corresponding Lévy–Khinchin spectral function by $H(x)$. Using the $A_2(\mathbf{E})$-property, we show that $\mathbf{E}$ contains a r.v. Y with the distribution F_H. Lemma 10 yields $S \in \mathbf{E}$.

Since the r.v.s Y_k are symmetric, the distribution of S depends on $|a_k|$ only. Hence, we may assume that

$$a_k > 0 \ (k = 1, 2, \dots) \quad , \quad \sum_{k=1}^{\infty} a_k^2 \leq 1. \tag{52}$$

The characteristic function $g(t)$ of Y_k is real. So,

$$g(t) = \exp\left(\int_{-\infty}^{\infty} (\cos(tx) - 1)\, dF(x)\right).$$

Lévy–Khinchin's formula has the form

$$g(t) = \exp\left(\int_{-\infty}^{\infty} (\cos(tx) - 1) \frac{1 + x^2}{x^2} dG(x)\right). \tag{53}$$

Comparing these formulae, we obtain

$$G(x) = \int_{-\infty}^{\infty} \frac{y^2}{1+y^2} dF(x). \tag{54}$$

Let $G_a(x)$ be the Lévy–Khintchin spectral function of the r.v. aY_k. Since $a_k > 0$ and the r.v. $a_k Y_k$ has the characteristic function $g(a_k t)$, the related Lévy–Khinchin spectral function is determined by the formula

$$G_{a_k}(x) = \int_{-\infty}^{x} \frac{a_k^2 + y^2}{1+y^2} dG\left(\frac{y}{a_k}\right). \tag{55}$$

Proposition 14. *The series*

$$H(x) = \sum_{k=1}^{\infty} G_{a_k}(x) \tag{56}$$

is uniformly convergent on **R** *and*

$$H(\infty) = \sum_{k=1}^{\infty} G_{a_k}(\infty).$$

Proof: Putting $u = y/a_k$, we obtain from (55)

$$\sum_{k=1}^{n} G_{a_k}(\infty) = \sum_{k=1}^{n} \int_{-\infty}^{\infty} \frac{a_k^2(1+u^2)}{1+a_k^2 u^2} dG(u) \leq \left(\sum_{k=1}^{n} a^2\right) \int_{-\infty}^{\infty} (1+u^2)\, dG(u)$$

for all integer n. Since

$$\int_{-\infty}^{\infty} x^2 dF(x) = EX_1^2 < \infty$$

then (54) implies that the integral in the right-hand side is finite, which with the inequality $0 \leq G_{a_k}(x) \leq G_{a_k}(\infty)$ gives us that

$$\sum_{k=1}^{\infty} G_{a_k}(\infty) < \infty.$$

Thus, the series (55) uniformly converges on **R**.

For every $x > 0$ and an integer n we have $H(x) \geq \sum_{k=1}^{n} G_{a_k}(x)$. Letting $x \to \infty$ and then $n \to \infty$, we obtain

$$H(\infty) \geq \sum_{k=1}^{\infty} G_{a_k}(\infty).$$

The converse estimate is obvious. □

We put for a bounded non-decreasing function $T(x)$ on $\mathbf{R}$ and $x > 0$

$$R(T, x) = (T(\infty) - T(x)) + (T(-x) - T(-\infty)). \tag{57}$$

Proposition 15. *Under the condition (52)*

$$R(H, x) \leq \sum_{k=1}^{\infty} R\left(F, \frac{x}{a_k}\right).$$

Proof: Proposition 14 gives us the equality

$$R(H, x) = \sum_{k=1}^{\infty} R(G_{a_k}, x).$$

From (52) $0 < a_k < 1$. Therefore, we have $(a_k^2 + y^2)/(1 + y^2) < 1$ and (55) yields $R(G_{a_k}, x) \leq R(G, x/a_k)$. From (54) $R(G, x) \leq R(F, x)$. The last bounds imply the needed relation. □

Now we may prove Theorem 7 for symmetric X_k. From (57) $R(F, x) \leq P\{|X_1| \geq x\}$, which together with (41) and Proposition 15 gives us the estimate $R(H, x) \leq Q_a(x)$, where $H(x)$ is determined by the formula (56). According to the $A_2(\mathbf{E})$-property, $\mathbf{E}$ contains a r.v. with the distribution function $H(x)/H(\infty)$. Since $H(x)$ is the Lévy–Khintchin spectral function corresponding to the sum (51), Lemma 10 pemits us to conclude that this sum belongs to $\mathbf{E}$. As mentioned above, this implies (40).

We turn to the general case. Put $Z_k = X_{2k} - X_{2k-1}$. Since $EX_1 = 0$, the inequality (1.13) holds. Thus, the estimate (40) for the r.v.s $\{X_k\}_{k=1}^{\infty}$ is equivalent to the same one for $\{Z_k\}_{k=1}^{\infty}$.

Let μX be the median of the r.v. X. From the well known symmetrization inequalities (see [20]) it follows that

$$\frac{1}{2} P\{|X_1| \geq x\} \leq P\{|Z_1| \geq x\} \leq 2P\left\{|X_1| \geq \frac{x}{2}\right\}$$

for $x \geq |\mu X|$. Thus, the conditions $X_1 \in A_2(\mathbf{E})$ and $Z_1 \in A_2(\mathbf{E})$ are equivalent. So, the general case is reduced to the symmetric one and Theorem 7 follows. □

3. Stable random variables with different exponents

1. Results. If $\{X_k\}_{k=1}^{\infty}$ are i.i.d.r.v.s with the symmetric q-stable distribution $(0 < q < 2)$, then the relation (1.21) holds. The question arises as to how (1.21) changes if the X_k have q-stable distributions with different exponents. Here we give the answer.

Let $0 < p < q_k < 2 \ (k \in \mathbf{N})$. We put for a real sequence $\mathbf{a} = \{a_k\}_{k=1}^{\infty}$

$$\Psi(\mathbf{a}) = \sum_{k=1}^{\infty} \frac{|a_k|^{q_k}}{q_k - p} \tag{58}$$

and

$$\beta(\mathbf{a}) = \inf \left\{ t > 0 : \Psi(t^{-1}\mathbf{a}) \leq 1 \right\}. \tag{59}$$

If $\Psi(\mathbf{a}) < \infty$, then the function $\Psi(t^{-1}\mathbf{a})$ is non-decreasing and continuous with respect to t. We have $\Psi(t^{-1}\mathbf{a}) \to \infty$ as $t \to 0$ and $\Psi(t^{-1}\mathbf{a}) \to 0$ as $t \to \infty$. Hence, the infimum in (59) is attained and

$$\Psi\left(\beta(\mathbf{a})^{-1}\mathbf{a}\right) = \sum_{k=1}^{\infty} \frac{|a_k|^{q_k}}{\beta(\mathbf{a})^{q_k}(q_k - p)} = 1. \tag{60}$$

If the exponents q_k are identical, then

$$\beta(\mathbf{a}) = \left(\sum_{k=1}^{\infty} \frac{|a_k|^{q_1}}{q_1 - p} \right)^{1/q_1}.$$

We say, as in [53], that a r.v. X has the *standard q-stable distribution* if in the formula (1.16) $\gamma = 1$.

Theorem 8. *Let $\{X_k\}_{k=1}^{\infty}$ be independent r.v.s with the standard q_k-stable distributions and let p be a fixed number such that $0 < p < q_k < 2$. Then for every finite collection $\mathbf{a} = \{a_k\}_{k=1}^{n}$ of real numbers*

$$A(p)\beta(\mathbf{a}) \leq \left(E \left| \sum_{k=1}^{n} a_k X_k \right|^p \right)^{1/p} \leq B(p)\beta(\mathbf{a}), \tag{61}$$

where

$$A(p) = \left(\pi^{-1}(1 - e^{-2})\Gamma(1 + p) \sin\left(\frac{\pi p}{2}\right) \right)^{1/p},$$

$$B(p) = \left(2\pi^{-1}(1 + p^{-1})\Gamma(1 + p) \sin\left(\frac{\pi p}{2}\right) \right)^{1/p}.$$

Applying this result we show that the estimate (37) cannot be improved.

Theorem 9. *Let $b_n > 0$, $b_n/n \searrow 0$ and $0 < p < 2$. There exist symmetric i.i.d.r.v.s $\{Y_k\}_{k=1}^{\infty}$ with finite pth absolute moment and such that*

$$E \left| \sum_{k=1}^{n} Y_k \right|^p \geq b_n$$

for all integers n.

2. Proof of Theorem 8. First we prove the right-hand side of (61). We use the following well known formula (see [55]). Let $f(t)$ be the characteristic function of the r.v. X and $E\,|X|^p < \infty$, where $0 < p < 2$. Then

$$E\,|X|^p = C(p) \int_0^\infty (1 - \mathrm{Re}f(t))\, t^{-p-1} dt,$$

where $C(p) = 2\pi^{-1}\Gamma(1+p)\sin(\pi p/2)$.

The sum $S = \beta(\mathbf{a})^{-1} \sum_{k=1}^n a_k X_k$ has the characteristic function

$$f(t) = \exp\left(-\sum_{k=1}^n \left| \frac{ta_k}{\beta(\mathbf{a})} \right|^{q_k} \right). \tag{62}$$

The last formula implies

$$E \left| \beta(\mathbf{a})^{-1} \sum_{k=1}^n a_k X_k \right|^p$$

$$= C(p) \int_0^\infty \left(1 - \exp\left(-\sum_{k=1}^n \left| \frac{ta_k}{\beta(\mathbf{a})} \right|^{q_k} \right) \right) t^{-p-1} dt. \tag{63}$$

We break up $(0, \infty)$ into the parts $(0, 1)$ and $[1, \infty)$. The difference in the brackets under the integral sign does not exceed 1. Hence, the integral over $[1, \infty)$ has the upper estimate $1/p$. For estimating the integral over $(0, 1)$ we use that $1 - \exp(-x) \le x$ for all $x \ge 0$. Applying (60), we find that this integral is not greater than

$$\int_0^1 \left(\sum_{k=1}^n \left| \frac{ta_k}{\beta(\mathbf{a})} \right|^{q_k} \right) t^{-p-1} dt = \sum_{k=1}^n \left| \frac{a_k}{\beta(\mathbf{a})} \right|^{q_k} \frac{1}{q_k - p} = 1.$$

So,

$$E \left| \beta(\mathbf{a})^{-1} \sum_{k=1}^n a_k X_k \right|^p \le C(p)(1 + p^{-1}),$$

which implies the right-hand side of (61).

We turn to the left-hand side. Since $0 < p < q_k < 2$, it follows that $1/(q_k - p) > 1/2$. Considering (60), we conclude that

$$\sum_{k=1}^n \left| \frac{ta_k}{\beta(\mathbf{a})} \right|^{q_k} \le 2 \sum_{k=1}^n \left| \frac{a_k}{\beta(\mathbf{a})} \right|^{q_k} \frac{1}{q_k - p} = 2$$

for $0 < t < 1$. It is not hard to verify $1 - \exp(-x) \geq \lambda x$ for $0 < x < 2$, where $\lambda = (1 - \exp(-2))/2$. Using (60) again, we get

$$E \left| \beta(\mathbf{a})^{-1} \sum_{k=1}^{n} a_k X_k \right|^p \geq \lambda C(p) \int_0^1 \left(\sum_{k=1}^{n} \left| \frac{ta_k}{\beta(\mathbf{a})} \right|^{q_k} \right) t^{-p-1} dt$$

$$= \lambda C(p) \sum_{k=1}^{n} \left| \frac{a_k}{\beta(\mathbf{a})} \right|^{q_k} \frac{1}{q_k - p} = \lambda C(p).$$

The last relations lead to the needed inequality. $\square$

3. On infinite sums. Here we show that the inequalities (61) are true for infinite sums, a result which will be applied for the proof of Theorem 9.

Lemma 11. *Assume the conditions of Theorem 8 and suppose that the sequence* $\mathbf{a} = \{a_k\}_{k=1}^{\infty}$ *is such that* $\Psi(\mathbf{a}) < \infty$. *Then the series* $\sum_{k=1}^{\infty} a_k X_k$ *converges almost surely, and the estimate (61) holds for it.*

Proof: First of all we show that the series under consideration is weakly convergent. The characteristic function $f_n(t)$ of the sum

$$S_n = \beta(\mathbf{a})^{-1} \sum_{k=1}^{n} a_k X_k$$

is determined by the formula (62). Since $0 < p < q_k < 2$, then $|t|^{q_k} \leq \max\{1, t^2\}$ and $1/(q_k - p) > 1/2$. This, (58) and the assumption $\Psi(\mathbf{a}) < \infty$ imply $\sum_{k=1}^{\infty} |a_k t|^{q_k} < \infty$ for all $t \in \mathbf{R}$. Consequently,

$$\lim_{n \to \infty} f_n(t) = \exp \left(-\sum_{k=1}^{\infty} \left| \frac{ta_k}{\beta(\mathbf{a})} \right|^{q_k} \right) \equiv f(t),$$

which implies the needed assertion (see [35]).

From the assumptions of Theorem 8, $|t|^{q_k} \geq |t|^p$ if $|t| \geq 1$. Using (60), we obtain

$$f(t) \leq \exp \left(-|t|^p \sum_{k=1}^{\infty} \left| \frac{a_k}{\beta(\mathbf{a})} \right|^{q_k} \right) = \exp \left(-|t|^p \right) \quad (|t| > 1).$$

So, $f(t)$ is integrable on $\mathbf{R}$ and the corresponding distribution is absolutely continuous [36]. According to the well known results of the equivalence of different forms of convergence, the considered series is convergent almost surely [53].

The inequality (61) for the partial sums yields the same estimate for the series. $\square$

4. Proof of Theorem 9. Let $\{X_k\}_{k=1}^{\infty}$ be a sequence of independent standard q_k-stable r.v.s and $0 < p < q_k < 2$. Consider independent r.v.s $\{X_{j,k}\}_{j,k=1}^{\infty}$ such that

$$X_{j,k} \overset{d}{=} X_k. \tag{64}$$

Let $\mathbf{a} = \{a_k\}_{k=1}^{\infty}$, $a_k > 0$ and $\Psi(\mathbf{a}) < \infty$. Define

$$Y_j = \sum_{k=1}^{\infty} a_k X_{j,k}. \tag{65}$$

According to Lemma 11 this series is convergent almost surely. The r.v.s $\{Y_j\}_{j=1}^{\infty}$ are symmetric, independent and identically distributed. Lemma 11 and the condition $\Psi(\mathbf{a}) < \infty$ imply $E\,|Y_j|^p < \infty$. The numbers a_k and q_k will be chosen below.

From (64) and (1.16)

$$\sum_{j=1}^{n} X_{j,k} \overset{d}{=} n^{1/q_k} X_k.$$

Using (65), we get

$$\sum_{j=1}^{n} Y_j = \sum_{k=1}^{\infty} a_k \sum_{j=1}^{n} X_{j,k} \overset{d}{=} \sum_{k=1}^{\infty} a_k n^{1/q_k} X_k.$$

Let

$$\beta_n = \beta\left(\left\{a_k n^{1/q_k}\right\}_{k=1}^{\infty}\right). \tag{66}$$

Lemma 11 leads to the estimate

$$E\left|\sum_{j=1}^{n} Y_j\right|^p = E\left|\sum_{k=1}^{\infty} a_k n^{1/q_k} X_k\right|^p \geq (A(p)\beta_n)^p. \tag{67}$$

We write (60) for the sequence $\left\{a_k n^{1/q_k}\right\}_{k=1}^{\infty}$ and get, using (66),

$$\sum_{k=1}^{\infty} \left(\frac{a_k}{\beta_n}\right)^{q_k} \frac{1}{q_k - p} = \frac{1}{n}.$$

Denote

$$v_k = \frac{a_k^{q_k}}{q_k - p} \quad , \quad h(t) = \sum_{k=1}^{\infty} v_k t^{-q_k}.$$

Then $h(\beta_n) = n^{-1}$.

The function $h(t)$ is decreasing. So, the estimate $(A(p)\beta_n)^p \geq b_n$ is equivalent to the condition

$$h\left(\frac{b_n^{1/p}}{A(p)}\right) \geq h(\beta_n) = \frac{1}{n}. \tag{69}$$

Thus, the problem is reduced to the following: given a sequence b_n, $b_n/n \searrow 0$ as $n \to \infty$, select numbers a_k and q_k such that (69) holds.

Let

$$\epsilon_n = \frac{b_n}{A(p)^p n}. \tag{70}$$

According to the conditions of the theorem, $\epsilon_n \searrow 0$. Put

$$v_k = 2(\epsilon_k - \epsilon_{k+1}) \tag{71}$$

and let $q_k \searrow p$. From (68), (70) and (71)

$$h\left(\frac{b_n^{1/p}}{A(p)}\right) = h\left((n\epsilon_n)^{1/p}\right) = 2\sum_{k=1}^{\infty}(\epsilon_k - \epsilon_{k+1})(n\epsilon_n)^{-q_k/p}.$$

It can be assumed that $\epsilon_n < 1$ for all n. Since $q_k \searrow p$, then $n^{-q_k/p} > n^{-q_n/p}$ for $k > n$ and $\epsilon_n^{-q_k/p} > \epsilon_n^{-1}$ for $k \geq 1$. From here and the above,

$$h\left(\frac{b_n^{1/p}}{A(p)}\right) \geq 2\sum_{k=n}^{\infty}(\epsilon_k - \epsilon_{k+1})(n\epsilon_n)^{-q_k/p}$$

$$\geq 2n^{-q_k/p}\epsilon_n^{-1}\sum_{k=n}^{\infty}(\epsilon_k - \epsilon_{k+1}) = 2n^{-q_k/n}. \tag{72}$$

We have used the equality $\sum_{k=n}^{\infty}(\epsilon_k - \epsilon_{k+1}) = \epsilon_n^{-1}$, which follows from the condition $\epsilon_n \searrow 0$.

We choose q_n so that

$$2n^{-q_n/p} = \frac{1}{n}. \tag{73}$$

From here

$$q_n = \frac{p(1 + \log 2)}{\log(n)}.$$

It is clear that $q_n \searrow p$. We define ϵ_n by the formula (70). According to (71) and (68), v_k and a_k are thereby uniquely determined. By (58) and (71)

$$\Psi(a) = \sum_{k=1}^{\infty}v_k = 2\sum_{k=1}^{\infty}(\epsilon_k - \epsilon_{k+1}) = \epsilon_1^{-1} < \infty,$$

where $\mathbf{a} = \{a_k\}_{k=1}^{\infty}$. The relations (72) and (73) imply (69) and the estimate $(A(p)\beta_n)^p \geq b_n$ for all n. Using (67), we get Theorem 9. $\square$

4. Equidistributed random variables in exponential Orlicz spaces

1. Results. Let $\{U_k\}_{k=1}^{\infty}$ be independent r.v.s. with the symmetric Bernoulli distribution, i.e. $P\{U_k = 1\} = P\{U_k = -1\} = 1/2$. Rodin and Semenov [48] proved that the norm $\|\sum_{k=1}^{n} a_k U_k\|_{(p)}$ is equivalent to $\|\{a_k\}\|_2$ if $1 \leq p \leq 2$ and to $\|\{a_k\}\|_{p',\infty}$ if $p > 2$ (the definition of the Lorentz sequence space $l_{r,s}$ see in the section 1.7). It was mentioned in the section 2 that in the case $1 \leq p \leq 2$ this relation holds for every sequence of i.i.d.r.v.s $\{X_k\}_{k=1}^{\infty} \subset L_{(p)}(\Omega)$. But it is not true in the case $p > 2$ (see Theorem 2.10). Here we investigate the conditions under which such a relation takes place for $p > 2$, i.e. we study the inequality

$$C_1 \|\mathbf{a}\|_{p',\infty} \leq \left\| \sum_{k=1}^{n} a_k X_k \right\|_{(p)} \leq C_2 \|\mathbf{a}\|_{p',\infty} , \qquad (74)$$

where $\mathbf{a} = \{a_k\}_{k=1}^{n}$ and C_1 and C_2 are positive and indepding of n and $\mathbf{a}$.

Theorem 10. *Let $p > 2$. The lower bound in (74) takes place for every sequence of i.i.d.r.v.s $\{X_k\}_{k=1}^{\infty} \subset L_{(p)}(\Omega)$ with mean zero, where*

$$C_1 = B(p) \|X_1\|_p$$

and $B(p)$ is a positive constant.

Let $f(t)$ be the characteristic function of the r.v. $X \in L_{(p)}(\Omega)$. As was mentioned in the section 2.4, $f(t)$ is extended to a entire function. The function

$$h(t) \equiv \log\left(f(-it)\right) = \log\left(E \exp\left(tX\right)\right) . \qquad (75)$$

plays an important role in our considerations.

Theorem 11. *Let $p > 2$, $\{X_k\}_{k=1}^{\infty} \subset L_{(p)}(\Omega)$ be i.i.d.r.v.s with mean zero and $h(t)$ be the related function determined by the formula (75). The upper estimate in (74) holds if and only if*

$$A(h) \equiv \int_0^{\infty} h(t) t^{-p'-1} dt < \infty . \qquad (76)$$

In this case $C_2 = D(p)A(p)^{1/p}$, where $D(p)$ is a constant.

We may formulate this result using the distribution of X_k directly. For this we need to recall some definitions.

We write $f(x) \approx g(x)$ $(x \to x_0)$ if there are positive constants a and b and a neighbouhood V of x_0 such that $f(ax) \leq g(x) \leq f(bx)$ for all $x \in V$.

Suppose $\phi(x)$ is positive function on $(0, \infty)$ such that $\phi(0) = 0$ and $\phi(x)/x \to \infty$ as $x \to \infty$. Put for real t

$$\psi(t) = \sup\{|tx| - \phi(x) : x > 0\}. \tag{77}$$

This function is said to be *complemented* to $\phi(x)$ (see [**28**], Ch.1). It may be easy verified that $\psi(t)/t \to \infty$ as $t \to \infty$ and that $\psi(t)$ is even and increasing on $(0, \infty)$.

Suppose $X \in L_{(p)}(\Omega)$ and $P\{|X| \geq x\} \neq 0$ for all positive x. Then the function $\phi(x) = -\log(P\{|X| \geq x\})$ is determine on $(0, \infty)$ and $\phi(x) \geq Cx^p$ for all $x > 0$ and some positive constant C. Therefore, if $p > 1$, there exists the corresponding complemented function $\psi(t)$. Using the inequality $|tx| \geq \phi(x) + \psi(t)$, which follows directly from (77), one may easily prove the next statement.

Proposition 16. *Let $P\{|X| \geq x\} \neq 0$ for all positive x and*

$$\phi(x) = -\log\left(P\{|X| \geq x\}\right).$$

Suppose $\phi(x)/x \to \infty$ as $x \to \infty$ and $\psi(t)$ is the corresponding complemented function. Then $h(t) \approx \psi(t)$ as $t \to \infty$.

So, condition (76) for $h(t)$ is equivalent to the same one for $\psi(t)$.

Now we describe some cases where (76) holds. If $|X| \leq C$, then $h(t) \approx |t|$ $(t \to \infty)$ (see [**34**], Ch.2). and (76) is fulfilled.

Suppose $\phi(x) = x^q$ and $q > p$. Then $\psi(t) = c|t|^{q'}$, where c is a positive constant. Since $q' < p'$, then (76) holds.

Let $\phi(x) = x^p \log^\gamma x$ for $x > 1$, where $\gamma > 0$ is a constant. Then $\psi(t) = t^{p'} \log^{-\gamma(p'-1)} t$ for t sufficiently big (see [**28**], Ch.1) and (76) is equivalent to the condition $\gamma > 1/(p'-1)$.

Using Kwapien—Richlick's inequality (see the section 1.4), Theorem 2.10 and Theorem 10 we conclude that for every sequence of i.i.d.r.v.s $\{X_k\}_{k=1}^\infty \subset L_{(p)}(\Omega)$ $(p > 2)$ with mean zero the estimate

$$D_1 \|a\|_{p',\infty} \leq \left\| \sum_{k=1}^n a_k X_k \right\|_{(p)} \leq D_2 \|a\|_{p'}$$

holds. Here positive D_1 and D_2 are independent of n and a_k.

2. Proof of Theorem 10. First we establish some auxiliary statemants.

Proposition 17. *Suppose that* $E \exp(tX) \le \exp\left(ct^{p'}\right)$ *for all* $t > 0$. *Then*

$$P\{X \ge x\} \le \exp\left(-\frac{b(p)x^p}{c^{p-1}}\right)$$

for all $x > 0$, *where* $b(p) = (p' - 1)/(p')^p$

Proof: We have $E \exp(tX) \ge \exp(tx)P\{X \ge x\}$ for all $x > 0$, which yields $P\{X \ge x\} \le \exp\left(ct^{p'} - tx\right)$. Taking the infimum of the right-hand side with respect to t we obtain the desired bound. $\square$

The following statement may easily be proved.

Proposition 18. *Suppose* $P\{X \ge x\} \le \exp(-Bx^p)$ *for all* $x > 0$ *and* $EX = 0$. *Then* $\|X\|_{(p)} \to 0$ *as* $B \to \infty$.

Let's denote by $h_X(t)$ the function determined by the formula (75). The next proposition follows directly from the previous ones.

Proposition 19. *Let* $p > 2$. *The following relation holds:*

$$d(p) = \inf\left\{\sup_{t \ne 0} \frac{h_X(t)}{|t|^{p'}} : \|X\|_{(p)} = 1,\, EX = 0\right\} > 0.$$

Now we may prove Theorem 10. Without loss of generality we may assume X_k to be symmetric, $\|X_k\|_{(p)} = 1$ and $\|\sum_{k=1}^{n} a_k X_k\|_{(p)} = 1$. Lemma 2.5 and (75) yield

$$\sum_{k=1}^{n} h(ta_k) \le C(p) \min\left\{t^2,\, |t|^{p'}\right\}. \tag{78}$$

Fixing $j \le n$ and putting $t = s/a_j^*$ we obtain

$$\sum_{k=1}^{j} h\left(\frac{sa_k^*}{a_j^*}\right) \le C(p)\left(a_j^*\right)^{-p'} s^{p'}.$$

One may easily verify that $h(t)$ is increasing on $(0, \infty)$. Hence the last sum is not less then $jh(s)$, which imlpies that $jh(s)\left(a_j^*\right)^{p'} \le C(p)s^{p'}$. Proposition 19 gives us

$$\|a\|_{p',\infty} \le \left(\frac{C(p)}{d(p)}\right)^{1/p}.$$

So, the lower estimate in (74) follows with the needed constant. $\square$

3. Proof of Theorem 11. As above we may assume X_k to be symmetric. Put for $\mathbf{a} = \{a_k\}_{k=1}^n$

$$\rho_h(\mathbf{a}) = \sup_{t \neq 0} |t|^{-1} \left(\sum_{k=1}^n h(a_k t) \right)^{1/p'}. \tag{79}$$

Since $p > 2$, then $p' < 2$. Using Lemma 2.9 we conclude that $\rho_h(\mathbf{a}) < \infty$ for each finite $\mathbf{a}$.

Since X_1 is symmetric, then $h(t)$ is even. From here

$$\rho_h(\{a_k\}) = \rho(\{|a_k|\}) \quad , \quad \rho_h(v\mathbf{a}) = |v|\,\rho_h(\mathbf{a}), \tag{80}$$

where $v \in \mathbf{R}$. As was mentioned above $h(t)$ is increasing on $(0, \infty)$, which yields that if $|a_k| \le |b_k|$, $k = 1, \ldots, n$, then $\rho_h(\mathbf{a}) \le \rho_h(\mathbf{b})$.

Lemma 12. *Let $p > 2$. There are positive constants $C = C(p)$ such that*

$$C^{-1} \rho_h(\mathbf{a}) \le \left\| \sum_{k=1}^n a_k X_k \right\|_{(p)} \le C \rho_h(\mathbf{a})$$

for each sequence $\{X_k\}_{k=1}^\infty \subset L_{(p)}(\Omega)$ of symmetric i.i.d.r.v.s and all $\mathbf{a} = \{a_k\}_{k=1}^n$.

Proof: From (79) $\sum_{k=1}^n h(a_k t) \le (|t|\,\rho_h(\mathbf{a}))^{p'}$. This and Lemma 2.7 imply the upper bound.

Turning to the lower one. Assume that $\|\sum_{k=1}^n a_k X_k\|_{(p)} = 1$. Then (78) takes place, which yields $\rho_h(\mathbf{a}) \le C(p)^{1/p'}$. Taking into account (80), we get the needed estimate and complete the proof. $\square$

Lemma 13. *Let $h(t)$ be an even positive function on $\mathbf{R}$, increasing on $(0, \infty)$. Let $r > 0$ and ρ_h be determined by the formula (79), where the exponent p' be substituted by r. Then the following conditions are equivalent:*

(i) $\rho_h(\mathbf{a}) \le C \|\mathbf{a}\|_{r,\infty}$ for all finite sequences $\mathbf{a}$, where C is a constant independent on $\mathbf{a}$;

(ii) $B(h) \equiv \int_0^\infty h(y) y^{-r-1} dy < \infty$.

If (ii) holds, then $C = (2rB(h))^{1/r}$.

Proof: $(i) \Rightarrow (ii)$. Let $t > 0$ and $a_k = n^{-1/r}$, $k = 1, \ldots, n$. Then $\|\mathbf{a}\|_{r,\infty} = 1$ and $C^r \ge \rho_h^r(\mathbf{a}) \ge t^{-r} \sum_{k=1}^n h(tk^{-1/r})$. Since $h(t)$ is increasing for positive t, we get

$$\sum_{k=1}^n h(tk^{-1/r}) \ge \sum_{k=1}^n \int_k^{k+1} h(ts^{-1/r}) ds = \int_1^{n+1} h(ts^{-1/r}) ds.$$

Putting $y = ts^{-1/r}$, we get for all positive t and integers n

$$C^r \geq \int_{t(n+1)^{-1/r}}^{t} rh(y)y^{-r-1}dy,$$

which yields (ii).

$(ii) \Rightarrow (i)$. Without loss of generality $\|\mathbf{a}\|_{r,\infty} = 1$, which gives $a_k^* \leq k^{-1/r}$, $k = 1,\ldots,n$. If $k \leq s \leq k+1$, then $k^{-1/r} \geq s^{-1/r} \geq (k+1)^{-1/r} \geq (2k)^{-1/r}$. From here, as above,

$$S(t) \equiv \sum_{k=1}^{n} h(ta_k^*) \leq \int_{1}^{n+1} h(2^{1/r}ts^{-1/r})ds$$

for all $t > 0$. Substituting variables by letting $y = 2^{1/r}ts^{-1/r}$, we get $S(t) \leq 2rt^r B(h)$, which implies (i), where $C = (2rB(h))^{1/r}$. $\square$

Theorem 11 follows from Lemmas 12 and 13. $\square$

CHAPTER IV

COMPLEMENTABILITY OF SUBSPACES
GENERATED BY INDEPENDENT RANDOM VARIABLES

1. Subspaces generated by bounded random variables

1. Introduction and result. We recall that a subspace **A** of a Banach space **B** is said to be *complemented* if there is a bounded linear operator $T : \mathbf{B} \to \mathbf{A}$ such that $Tx = x$ for every $x \in \mathbf{A}$. The operator T is called a *projector*. The *closure subspace* in **B** generated by elements $\{x_k\}_{k=1}^{\infty} \subset \mathbf{B}$, is denoted by span $\{x_k\}_{k=1}^{\infty}$.

In this chapter we suppose that the probability space is the segment $[0, 1]$ with Lebesgue measure, i.e. all considered r.v.s are determined on this segment. Applying the well known theorems about isomorphism of the measure spaces, one may pass to the general case.

The main result of this section is the following.

Theorem 1. *Let* **E** *be a r.i. space and* $\{X_k\}_{k=1}^{\infty}$ *be independent bounded r.v.s such that*

$$0 < \inf_k \|X_k - EX_k\|_1 \le \sup_k \|X_k - EX_k\|_\infty < \infty. \tag{1}$$

Then the subspace span $\{X_k\}_{k=1}^{\infty} \subset \mathbf{E}$ *is complemented if and only if each of the spaces* $\mathbf{E}'$ *and* $\mathbf{E}''$ *contains a r.v. with the normal distribution.*

For the Rademacher system this result has been proved in [33] and [49].

First we show that, without loss of generality, we may assume $EX_k = 0$ ($k \in \mathbf{N}$). Put $Y_k = X_k - EX_k$ and

$$L = \operatorname{span}\left\{ \{X_k\}_{k=1}^{\infty} \bigcup \{I_{[0,1]}\} \right\} \subset \mathbf{E}.$$

The subspaces span $\{X_k\}_{k=1}^{\infty}$ and span $\{Y_k\}_{k=1}^{\infty}$ belong to L and have co-dimension not greater than 1. So, these subspaces are simultaneously complemented or not.

2. Proof of Theorem 1. Sufficiency. Since $EX_k = 0$, then (1) yields

$$0 < \inf_k \left\{ EX_k^2 \right\} \le \sup_k \left\{ EX_k^2 \right\} < \infty,$$

which implies the r.v.s $\left(EX_k^2 \right)^{-1/2} X_k$ also satisfy (1). So, we may assume that $EX_k^2 = 1$ for all $k \in \mathbf{N}$.

Denote

$$\langle X , Y \rangle = E(XY). \tag{2}$$

We have

$$\langle X_j , X_k \rangle = \delta_{j,k}, \tag{3}$$

which yields that the operator $Q_n X = \sum_{k=1}^{n} \langle X , X_k \rangle X_k$ is a projector on the subspace span $\{X_k\}_{k=1}^{n}$. First we show the uniform boundedness of the progectors $\{Q_n\}_{n=1}^{\infty}$.

It is easy to verify that from (1) the conditions for Paley—Zygmund's and Bernstein's inequalities follow (see the section 1.4). As mentioned in the section 1.1, if X is bounded, then $\|X\|_{\mathbf{E}} = \|X\|_{\mathbf{E}''}$. Since each of the spaces $\mathbf{E}'$ and $\mathbf{E}''$ contains a r.v. with the normal distribution, there are positive constants C and C_1 such that

$$C^{-1} \left(\sum_{k=1}^{n} a_k^2 \right)^{1/2} \leq \left\| \sum_{k=1}^{n} a_k X_k \right\|_{\mathbf{E}} \leq C \left(\sum_{k=1}^{n} a_k^2 \right)^{1/2} ,$$

$$\tag{4}$$

$$C_1^{-1} \left(\sum_{k=1}^{n} b_k^2 \right)^{1/2} \leq \left\| \sum_{k=1}^{n} a_k X_k \right\|_{\mathbf{E}'} \leq C_1 \left(\sum_{k=1}^{n} b_k^2 \right)^{1/2}$$

for all real a_k and b_k and all integers n. From here and (3)

$$\left\| \sum_{k=1}^{n} a_k X_k \right\|_{\mathbf{E}} \leq C \left(\sum_{k=1}^{n} a_k^2 \right)^{1/2}$$

$$= C \sup \left\{ \left\langle \sum_{k=1}^{n} a_k X_k , \sum_{k=1}^{n} b_k X_k \right\rangle : \sum_{k=1}^{n} b_k^2 \leq 1 \right\}$$

$$\leq C \left\{ \left\langle \sum_{k=1}^{n} a_k X_k , Y \right\rangle : Y \in \text{span} \{X_k\}_{k=1}^{n} , \|Y\|_{\mathbf{E}'} \leq C_1 \right\}.$$

Since $\langle Q_n X, Y \rangle = \langle X, Q_n Y \rangle$, then

$$\|Q_n X\|_{\mathbf{E}} \leq C C_1 \sup \{\langle X, Q_n Y \rangle : Y \in \text{span} \{X_k\}_{k=1}^{n} , \|Y\|_{\mathbf{E}'} \leq 1\}.$$

If $Y \in \text{span} \{X_k\}_{k=1}^{n}$, then $Q_n Y = Y$ and we have

$$\|Q_n X\|_{\mathbf{E}} \leq C C_1 \sup \{\langle X, Y \rangle : Y \in \mathbf{E}', \|Y\|_{\mathbf{E}'} \leq 1\} \leq C C_1 \|X\|_{\mathbf{E}}. \tag{5}$$

Let's put

$$QX = \sum_{k=1}^{\infty} \langle X, X_k \rangle X_k.$$

We show that the series converges in $\mathbf{E}$ for each $X \in \mathbf{E}$. According to (4) and (5)

$$C^{-1} \left(\sum_{k=1}^{n} \langle X, X_k \rangle^2 \right)^{1/2} \leq \|Q_n X\|_{\mathbf{E}} \leq CC_1 \|X\|_{\mathbf{E}}.$$

Hence $\sum_{k=1}^{\infty} \langle X, X_k \rangle^2 \leq C^4 C_1^2 \|X\|_{\mathbf{E}}^2 < \infty$. Using (4) once more, we get for all $m < n$

$$\left\| \sum_{k=m}^{n} \langle X, X_k \rangle X_k \right\|_{\mathbf{E}} \leq C \left(\sum_{k=m}^{n} \langle X, X_k \rangle^2 \right)^{1/2}.$$

From here and the above, the convergence of the series follows.

So, the operator Q is defined on $\mathbf{E}$. It is obvious that Q is a projector on the subspace span $\{X_k\}_{k=1}^{\infty}$. According to (5)

$$\|QX\|_{\mathbf{E}} = \lim_{n \to \infty} \|Q_n X\|_{\mathbf{E}} \leq CC_1 \|X\|_{\mathbf{E}}$$

and sufficiency is proved.

3. Outline proof of the necessity. The proof of the necessity is very bulky and therefore we describe it in outline. First we reduce the proof to the case of separable $\mathbf{E}$ and symmetric X_k. In such a space the set of all finite-valued r.v.s is dense, which permits us to assume that X_k is finite-valued. For every collection $\{X_k\}_{k=1}^{n}$ $(n \in \mathbf{N})$ we construct an equidistributed collection $\{\bar{X}_k\}_{k=1}^{n}$ such that each set of the type $\{\bar{X}_k = \text{const.}\}$ is a finite union of mutually disjoint intervals. Then applying the assumption of complementability of the subspace span $\{X_k\}_{k=1}^{\infty}$, we show that there is a sequence of uniformly bounded projectors on the subspaces span $\{\bar{X}_k\}_{k=1}^{n}$ $(n \in \mathbf{N})$. The simple structure of the r.v.s $\bar{X}_k$ permits us to determine for each n a group of measure-preserving transformations ϕ on the interval $[0, 1]$ such that each operator $TX(t) = X(\phi(t))$ maps the subspace span $\{\bar{X}_k\}_{k=1}^{n}$ into itself. Averaging, we construct, as in [49], the projector $Q_n : \mathbf{E} \to \text{span} \{\bar{X}_k\}_{k=1}^{n}$ commuting with the action of the group. The projectors Q_n are uniformly bounded, which permits us to construct a system of uniformly bounded projectors on the subspaces span$\{r_k\}_{k=1}^{n}$ $(n \in \mathbf{N})$, where

$$r_k(t) = \text{sign} \left(\sin \left(2^k \pi t \right) \right)$$

are Rademacher functions. Applying the results of [33] and [49], we complete the proof.

4. Symmetrization. The next statement reduces our consideration to the case of symmetric r.v.s.

Lemma 1. *Let an r.i. space* **E** *be separable or maximal and* $\{X_k\}_{k=1}^{\infty} \subset \mathbf{E}$ *be independent r.v.s. There are independent symmetric r.v.s* $\{Y_k\}_{k=1}^{\infty}$ *such that*

    1) the subspaces span $\{X_k\}_{k=1}^{\infty} \subset \mathbf{E}$ *and span* $\{Y_k\}_{k=1}^{\infty}$ *are isomorphic;*

    2) if the subspace span $\{X_k\}_{k=1}^{\infty}$ *is complemented in* **E***, then the subspace* span $\{Y_k\}_{k=1}^{\infty}$ *has the same property;*

    3) if the sequence $\{X_k\}_{k=1}^{\infty}$ *satisfies (1), then* $\{Y_k\}_{k=1}^{\infty}$ *also satisfies (1);*

    4) if $\{X_k\}_{k=1}^{\infty}$ *are equidistributed r.v.s, then so are* $\{Y_k\}_{k=1}^{\infty}$.

Proof: Let $\left\{\bar{X}_k\right\}_{k=1}^{\infty}$ be an independent copy of the sequence $\{X_k\}_{k=1}^{\infty}$ and $Y_k = X_k - \bar{X}_k$. As mentioned above, we may assume $EX_k = 0$. We put $L = \mathrm{span}\,\{X_k\}_{k=1}^{\infty}$, $N = \mathrm{span}\,\{Y_k\}_{k=1}^{\infty}$ and define the operator A from L to N by the formula $AX_k = Y_k$ $(k = 1, 2, \ldots)$. According to Proposition 1.11 and the remark following it,

$$\|X\|_{\mathbf{E}} \leq \|AX\|_{\mathbf{E}} \leq 2\,\|X\|_{\mathbf{E}} \qquad (6)$$

for every $X \in L$. Hence, A is an isomorphism.

    Suppose L is complemented in **E** and denote the corresponding projector by Q. Let β be the σ-algebra generated by the r.v.s $\{X_k\}_{k=1}^{\infty}$. Since **E** is separable or maximal, then the operator E^{β} of the conditional expectation is bounded in **E** and $\left\|E^{\beta}\right\|_{\mathbf{E}\to\mathbf{E}} = 1$. Putting $Q_1 = AQE^{\beta}$ we have $E^{\beta}Y_k = E^{\beta}(X_k - \bar{X}_k) = X_k - E\bar{X}_k = X_k$, which implies that $Q_1 Y_k = AQ_1 X_k = AX_k = Y_k$. Taking into account (6), we conclude that Q is a bounded projector on N.

    The third assertion of the lemma follows from Proposition 1.11 applied to $L_1(0,1)$ and $L_{\infty}(0,1)$. The last assertion is obvious. $\square$

5. Reduction to separable spaces. Let's recall that $\mathbf{E}_0$ is the closure of $L_{\infty}(0,1)$ in the r.i. space **E**. If $\mathbf{E} \neq L_{\infty}(0,1)$, then $\mathbf{E}_0$ is a separable r.i. space (see [48]). From the conditions of Theorem 1, $X_k \in \mathbf{E}_0$. The complementability of span $\{X_k\}_{k=1}^{\infty}$ in **E** implies the same in $\mathbf{E}_0$. Since $(\mathbf{E}_0)' = \mathbf{E}'$, then, if $\mathbf{E} \neq L_{\infty}(0,1)$, we may assume **E** to be separable. The next assertion permits to us to eliminate this assumption.

Proposition 1. *Each infinite dimentional subspace generated in* $L_{\infty}(0,1)$ *by independent r.v.s is non-complemented.*

Proof: Let $\{X_k\}_{k=1}^{\infty} \subset L_{\infty}(0,1)$ be independent r.v.s. According to Lemma 1 we may assume X_k to be symmetric. In addition, without loss of generality,

$$\|X_k\|_{\infty} = 1 \qquad (7)$$

for all $k \in \mathbf{N}$. First we show that the subspace span $\{X_k\}_{k=1}^{\infty} \subset L_{\infty}(0,1)$ is isomorphic to the space l_1. Consider a collection $\{a_k\}_{k=1}^{n}$ of real numbers and put $\epsilon_k = \text{sign}(a_k)$. Denoting the Lebesgue measure by μ, we have

$$\mu\left\{\epsilon_k X_k \geq \frac{1}{2} \ (1 \leq k \leq n)\right\} = \prod_{k=1}^{n} \mu\left\{\epsilon_k X_k \geq \frac{1}{2}\right\} > 0.$$

Hence, the strong inequality

$$\mu\left\{\left|\sum_{k=1}^{n} a_k X_k\right| \geq \frac{1}{2}\sum_{k=1}^{n}|a_k|\right\} \geq \mu\left\{\epsilon_k X_k \geq \frac{1}{2} \ (1 \leq k \leq n)\right\} > 0$$

holds and we get

$$\left\|\sum_{k=1}^{n} a_k X_k\right\|_{\infty} \geq \frac{1}{2}\sum_{k=1}^{n}|a_k|.$$

Besides,

$$\left\|\sum_{k=1}^{n} a_k X_k\right\|_{\infty} \leq \sum_{k=1}^{n}|a_k|\,\|X_k\|_{\infty} = \sum_{k=1}^{n}|a_k|.$$

These estimates imply the needed statement.

It was proved in [41], that the spaces $L_\infty(0,1)$ and l_∞ are isomorphic. According to [42] each complemented infinite dimentional subspace in l_∞ is not separable. Since the space l_1 is separable, the considered subspace span $\{X_k\}_{k=1}^{\infty}$ is non-complemented in $L_\infty(0,1)$. $\square$

6. The construction of invariant projections. Let n be fixed and $\{X_k\}_{k=1}^{n}$ be independent symmetric finite-valued r.v.s. Denote the positive values of the r.v. X_k in increasing order by $a^{(k)}(v_k)$ $(2 \leq v_k \leq m_k)$. We put

$$a^{(k)}(\pm 1) = 0 \ , \ a^{(k)}(-v_k) = -a^{(k)}(v_k) \ , \ \Theta_m = \{\pm 1, \ldots, \pm m\}.$$

Let's denote $\alpha^{(k)}(\pm 1) = \frac{1}{2}\mu\{X_k = 0\}$ and

$$\alpha^{(k)}(v_k) = \mu\left\{X_k = a^{(k)}(v_k)\right\} \quad (v_k \in \Theta_{m_k}, \ v_k \neq \pm 1). \tag{8}$$

Since X_k is symmetric, then

$$\alpha^{(k)}(-v_k) = \alpha^{(k)}(v_k). \tag{9}$$

The collection of integral n-dimensional vectors

$$U_n = \{\mathbf{v} = (v_1, \ldots, v_n) : v_k \in \Theta_{m_k} \ (1 \leq k \leq n)\} \tag{10}$$

plays an important role in our constructions.

Proposition 2. *There are mutually disjoint intervals* $\Delta(\mathbf{v})$ $(\mathbf{v} \in U_n)$ *such that*

1) $\mu(\Delta(\mathbf{v})) = \prod_{k=1}^{n} \alpha^{(k)}(v_k)$, $\mathbf{v} = (v_1, ..., v_n)$;
2) $\bigcup_{\mathbf{v} \in U_n} \Delta(\mathbf{v}) = (0, 1)$.

Proof: From the definition

$$\sum_{v_k \in \Theta_{m_k}} \alpha^{(k)}(v_k) = 1 \quad (1 \le k \le n), \tag{11}$$

which implies

$$\sum_{\mathbf{v} \in U_n} \prod_{k=1}^{n} \alpha^{(k)}(v_k) = \prod_{k=1}^{n} \left(\sum_{v_k \in \Theta_{m_k}} \alpha^{(k)}(v_k) \right) = 1.$$

So, there exist intervals with the needed properties. $\square$

We determine the r.v. $\bar{X}_k$ by the equality

$$\bar{X}_k(t) = a^{(k)}(v_k) , \ t \in \Delta(\mathbf{v}) , \ \mathbf{v} = (v_1, \dots, v_n). \tag{12}$$

The next assertion easily follows from the previous.

Proposition 3. *The r.v.s* $\left\{ \bar{X}_k \right\}_{k=1}^{n}$ *are independent and symmetric, and* $\bar{X}_k \stackrel{d}{=} X_k$ $(1 \le k \le n)$.

Now we reduce the problem to a study of the subspaces generated by the r.v.s $\bar{X}_k$.

Proposition 4. *Let* $\{X_k\}_{k=1}^{n}$ *and* $\left\{ \bar{X}_k \right\}_{k=1}^{n}$ *be equidistributed collections of finite-valued r.v.s. Then there is an inverse measure-preserving transformation* $\psi_n : [0,1] \to [0,1]$ *such that* $X_k(\psi_n(t)) = \bar{X}_k(t)$ $(1 \le k \le n)$.

Proof: Without loss of generality we may assume the r.v.s $\bar{X}_k$ and X_k to be Borel measurable. We denote the values taken by X_k by $b^{(k)}(j_k)$ $(1 \le j_k \le s_k)$. Let V_n be the collection of all integral vectors of the type $\mathbf{j} = (j_1, \dots, j_n)$ $(1 \le j_k \le s_k , 1 \le k \le n)$. Put

$$e_{\mathbf{j}} = \left\{ X_k = b^{(k)}(j_k), \ (1 \le k \le n) \right\}$$

and

$$\bar{e}_{\mathbf{j}} = \left\{ \bar{X}_k = b^{(k)}(j_k), \ (1 \le k \le n) \right\},$$

where $\mathbf{j} \in V_n$. We denote the Borel σ-algebra on $[0,1]$ by β. Let $\beta_{\mathbf{j}}$ be the collection of all sets of the type $e \bigcap e_{\mathbf{j}}$ $(e \in \beta)$ and let $\mu_{\mathbf{j}}$ be the restriction of μ on $\beta_{\mathbf{j}}$. The σ-algebra $\bar{\beta}_{\mathbf{j}}$ and the measure $\bar{\mu}_{\mathbf{j}}$ are determined similarly. Since

$\bar{X}_k \overset{d}{=} X_k$, then $\mu e_j = \mu \bar{e}_j$. From here and the well known theorem about isomorphism of σ-algebras the existence of an isomorphism $h_j : \bar{\beta}_j \to \beta_j$ follows for each $j \in V_n$.

The sets e_j are mutually disjoint and their union is a set of full measure and the sets $\bar{e}_j$ are the same. So, the isomorphisms h_j generate the isomorphism $h_n : \beta \to \beta$. It is well known that such an isomorphism is induced by a measure-preserving transformation $\psi_n : [0,1] \to [0,1]$.

According to the construction, $\psi_n(\bar{e}_j) = e_j$ for all $j \in V_n$, which implies the needed relations. $\square$

Proposition 5. *Let the condition of Proposition 4 hold, $\mathbf{E}$ be a r.i. space and $Q_n : \mathbf{E} \to \mathrm{span}\{X_k\}_{k=1}^{n}$ be a projector. Then there exist a projector $\bar{Q}_n : \mathbf{E} \to \mathrm{span}\{\bar{X}_k\}_{k=1}^{n}$ such that*

$$\|\bar{Q}_n\|_{\mathbf{E} \to \mathbf{E}} \leq \|Q_n\|_{\mathbf{E} \to \mathbf{E}} .$$

Proof: We put $(T_{\psi_n} X)(t) = X(\psi_n(t))$ and

$$\bar{Q}_n = T_{\psi_n} Q T_{\psi_n}^{-1} .$$

It is not hard to verify that $\bar{Q}_n$ is the needed projector. $\square$

Denote by V^n the group of all linear operators in $\mathbf{R}^n$ changing the signs of some coordinates of vectors. More precisely, if $g \in V^n$, then there is a collection of numbers $\epsilon_k(g) = \pm 1$ such that

$$g(\mathbf{v}) = (\epsilon_1(g)v_1, \dots, \epsilon_n(g)v_n),$$

where $\mathbf{v} = (v_1, ..., v_n)$. It follows from (10) and the definition of Θ_m that $g(U_n) = U_n$ for all $g \in V^n$. According to (9) and Proposition 2, $\mu\Delta(g(\mathbf{v})) = \mu\Delta(\mathbf{v})$ for all $\mathbf{v} \in U_n$ and $g \in V^n$.

Let $\Delta(\mathbf{v}) = (\alpha, \beta)$ and $\Delta(g(\mathbf{v})) = (\gamma, \delta)$. Put

$$\phi_g(t) = \begin{cases} t + (\gamma - \alpha) & \text{if } t \in (\alpha, \beta), \\ t + (\alpha - \gamma) & \text{if } t \in (\gamma, \delta). \end{cases} \tag{13}$$

This equality and Proposition 2 imply that $\phi_g(t)$ is an inverse measure-preserving transformation on $[0,1]$. Moreover,

$$\phi_g(\Delta(\mathbf{v})) = \Delta(g(\mathbf{v})), \tag{14}$$

which yields $\phi_{g_1}\phi_{g_2} = \phi_{g_1 g_2}$.

We put for $g \in V^n$

$$T(g)X(t) = X(\phi_g(t)).$$

Since $T(g)X \overset{d}{=} X$, then $T(g)$ acts isometrically in every r.i. space. The mapping $g \to T(g)$ is the representation of the group V^n in such a space. We show that the subspace $\mathrm{span}\{\bar{X}_k\}_{k=1}^{n}$ is invariant with respect to the representation $T(g)$.

Proposition 6. *The equalities $T(g)\bar{X}_k = \epsilon_k(g)\bar{X}_k$ are valid.*

Proof: Let $\mathbf{v} \in U_n$. According to (14),

$$T(g)\bar{X}_k\Big|_{\Delta(\mathbf{v})} = \bar{X}_k\Big|_{\Delta(g(\mathbf{v}))}.$$

From the definition, $a^{(k)}(-v_k) = -a^{(k)}(v_k)$. This and (12) imply

$$\bar{X}_k\Big|_{\Delta(g(\mathbf{v}))} = a^{(k)}(\epsilon_k(g)v_k) = \epsilon_k(g)a^{(k)}(v_k) = \epsilon_k(g)\bar{X}_k\Big|_{\Delta(\mathbf{v})}.$$

Hence,

$$T(g)\bar{X}_k\Big|_{\Delta(\mathbf{v})} = \epsilon_k(g)\bar{X}_k\Big|_{\Delta(\mathbf{v})}.$$

Using Proposition 2 we get the needed equality. $\square$

The next step is to construct an invariant projector on the subspace $\mathrm{span}\left\{\bar{X}_k\right\}_{k=1}^{n} \subset \mathbf{E}$.

Lemma 2. *Let $P_n : \mathbf{E} \to \mathrm{span}\left\{\bar{X}_k\right\}_{k=1}^{n}$ be a projector. There is a collection $\left\{\bar{Y}_k\right\}_{k=1}^{n} \subset \mathbf{E}'$ such that*
1) $\left\langle \bar{X}_j , \bar{Y}_k \right\rangle = \delta_{j,k}$;
2) the r.v.s $\bar{Y}_k$ are constant on each interval $\Delta(\mathbf{v})$, $\mathbf{v} \in U_n$;
3) the projector

$$Q_n X = \sum_{k=1}^{n} \left\langle X , \bar{Y}_k \right\rangle \bar{X}_k \tag{15}$$

commutes with every operator $T(g)$, $g \in V^n$;
4) $\|Q_n\|_{\mathbf{E}\to\mathbf{E}} \le \|P_n\|_{\mathbf{E}\to\mathbf{E}}$.

Proof: Since $\mathbf{E}$ is separable, then $\mathbf{E}' = \mathbf{E}^*$, where $\mathbf{E}^*$ is the conjugate space. Using Proposition 1.14, we conclude that there is a collection $\{Z_k\}_{k=1}^{n} \subset \mathbf{E}'$ such that

$$P_n X = \sum_{k=1}^{n} \left\langle X , Z_k \right\rangle \bar{X}_k .$$

Let β be the σ-algebra generated by all the intervals $\Delta(\mathbf{v})$, $\mathbf{v} \in U_n$. According to the definition, the r.v.s $\bar{X}_k$ $(1 \le k \le n)$ are β-measurable. Hence, $E^\beta \bar{X}_k = \bar{X}_k$ and the operator $\bar{P}_n = E^\beta P_n E^\beta$ is a projector on the subspace $\mathrm{span}\left\{\bar{X}_k\right\}_{k=1}^{n}$. It is easy to verify that $\left\langle E^\beta X , Y \right\rangle = \left\langle X , E^\beta Y \right\rangle$. From here

$$\bar{P}_n X = \sum_{k=1}^{n} \left\langle X , E^\beta Z_k \right\rangle \bar{X}_k .$$

It is clear that each r.v. $\bar{Z}_k = E^\beta Z_k$ is a constant on $\Delta(\mathbf{v})$, $\mathbf{v} \in U_n$.

Since $\mathbf{E}$ is separable, the operator E^β is bounded on $\mathbf{E}$ and has the norm 1 (see Proposition 1.4). Hence, $\left\| \bar{P}_n \right\|_{\mathbf{E}\to\mathbf{E}} \leq \|P_n\|_{\mathbf{E}\to\mathbf{E}}$. We put

$$Q_n X = 2^{-n} \sum_{g\in V^n} \left(T(g)\bar{P}_n T(g^{-1}) \right) X.$$

As mentioned above, $T(g)$ acts isometrically in $\mathbf{E}$. Since the group V^n consists of 2^n elements, then the estimate $\|Q_n\|_{\mathbf{E}\to\mathbf{E}} \leq \left\| \bar{P}_n \right\|_{\mathbf{E}\to\mathbf{E}}$ holds. It is not hard to verify that $T(g)Q_n = Q_n T(g)$ for each $g \in V^n$. Proposition 6 permits us to conclude that Q_n is a projector on span $\left\{ \bar{X}_k \right\}_{k=1}^n$.

Since

$$\langle T\left(g^{-1}\right) X, Y \rangle = \langle X, T(g)Y \rangle \tag{16}$$

for every $g \in V^n$, then using Proposition 6, we get

$$Q_n X = \sum_{k=1}^n \left(2^{-n} \sum_{g\in V^n} \langle X, T(g)Z_k \rangle \, \epsilon_k(g) \right) \bar{X}_k.$$

The r.v.s $T(g)\bar{Z}_k$ are constants on each of the segments $\Delta(\mathbf{v})$. Put

$$\bar{Y}_k = 2^{-n} \sum_{g\in V^n} \epsilon_k(g) T(g)\bar{Z}_k.$$

Then we get (15) and the statements 2) and 4) of the lemma. Since Q_n is a projector, then 1) holds. $\square$

It is obvious that $g^{-1} = g$ for all $g \in V^n$, which yields that $T(g^{-1}) = T(g) = (T(g))^{-1}$. This, (16), Proposition 6 and the permutability of Q_n and the operators $T(g)$ imply that

$$T(g)\bar{Y}_k = \epsilon_k(g)\bar{Y}_k \quad (1 \leq k \leq n, g \in V^n). \tag{17}$$

7. Reduction to the Rademacher system. For $\mathbf{v} = (v_1,\ldots,v_n)$ and $t \in \Delta(\mathbf{v})$ we put

$$\bar{r}_k(t) = \begin{cases} \operatorname{sign}\left(\bar{Y}_k \big|_{\Delta(\mathbf{v})} \right) & \text{if } \; \bar{Y}_k \big|_{\Delta(\mathbf{v})} \neq 0, \\[2mm] \operatorname{sign}(v_k) & \text{if } \; \bar{Y}_k \big|_{\Delta(\mathbf{v})} = 0. \end{cases} \tag{18}$$

This definition is valid because $\bar{Y}_k$ is constant on each interval $\Delta(\mathbf{v})$.

Proposition 7. *For every* $g \in V^n$ *and* $k = 1, \ldots, n$

$$T(g)\bar{r}_k = \epsilon_k(g)\bar{r}_k .$$

Proof: We have

$$\left. (T(g)\bar{r}_k) \right|_{\Delta(v)} = \left. \bar{r}_k \right|_{\Delta(g(v))} .$$

According to (17)

$$\left. \bar{Y}_k \right|_{\Delta(g(v))} = \left. (T(g)\bar{Y}_k) \right|_{\Delta(v)} = \left. \epsilon_k(g)\bar{Y}_k \right|_{\Delta(v)} .$$

From here and (18) the needed equality follows. $\Box$

Proposition 8. *The r.v.s* $\{\bar{r}_k\}_{k=1}^n$ *are symmetric and independent.*

Proof: Let k, $1 \leq k \leq n$, be fixed and $g \in V^n$ be such that $\epsilon_k(g) = -1$. From Proposition 7, $T(g)\bar{r}_k = -\bar{r}_k$. Since ϕ_g is a measure-preserving transformation, it follows from the definition of $T(g)$ that

$$\bar{r}_k \overset{d}{=} -\bar{r}_k ,$$

i.e. $\bar{r}_k$ is symmetric.

Let $e(g) = \{\bar{r}_k = \epsilon_k(g), 1 \leq k \leq n\}$, where $g \in V^n$. The sets $e(g), g \in V^n$, are mutually disjoint and their union is $(0, 1)$. From Proposition 7

$$
\begin{aligned}
e(gg_1) &= \{\bar{r}_k = \epsilon_k(g)\epsilon_k(g_1), 1 \leq k \leq n\} \\
&= \{T(g)\bar{r}_k = \epsilon_k(g_1), 1 \leq k \leq n\} = \phi_g^{-1}(e(g_1)) .
\end{aligned}
$$

This relation implies $\mu e(g) = \mu e(g')$ for all g, $g' \in V^n$. Since the group V^n consists of 2^n elements, then $\mu e(g) = 2^{-n}$.

By virtue of symmetry, $\mu\{\bar{r}_k = \epsilon_k(g)\} = 1/2$. Each collection of numbers $\epsilon_k = \pm 1$ $(1 \leq k \leq n)$ determines an element $g \in V^n$. So,

$$\mu\{\bar{r}_k = \epsilon_k, 1 \leq k \leq n\} = \mu e(g) = 2^{-n} = \prod_{k=1}^n \mu\{\bar{r}_k = \epsilon_k\} .$$

Hence, the r.v.s $\{\bar{r}_k\}_{k=1}^n$ are independent. $\Box$

Proposition 9. *The following equalities are valid:*

$$\langle \bar{r}_k, \bar{Y}_j \rangle = \delta_{j,k} E\left|\bar{Y}_k\right| \quad (1 \leq j, k \leq n).$$

Proof: Let $j \neq k$ and choose $g \in V^n$ so that $\epsilon_j(g) = 1$ and $\epsilon_k(g) = -1$. As mentioned above, $T(g^{-1}) = T(g)$. Using (16), (17) and Proposition 7, we obtain

$$\langle \bar{r}_k , \bar{Y}_j \rangle = \langle \epsilon_k(g)T(g)\bar{r}_k , \bar{Y}_j \rangle = \epsilon_k(g) \langle \bar{r}_k , T(g)\bar{Y}_j \rangle$$
$$= \epsilon_k(g)\epsilon_j(g) \langle \bar{r}_k , \bar{Y}_j \rangle = - \langle \bar{r}_k , \bar{Y}_j \rangle .$$

Thus, for $j \neq k$ the equality is proved.

According to Lemma 2, the r.v. $\bar{Y}_k$ is a constant on each interval $\Delta(\mathbf{v})$. Using (18), we get

$$\langle \bar{r}_k , \bar{Y}_j \rangle = \sum_{\mathbf{v} \in U_n} \left| \bar{Y}_k \Big|_{\Delta(\mathbf{v})} \right| \mu\Delta(\mathbf{v}) = E \left| \bar{Y}_k \right| . \square$$

We put

$$\tilde{Q}_n X = \sum_{k=1}^n \frac{\langle X , \bar{Y}_k \rangle}{(E |\bar{Y}_k|)} \bar{r}_k . \tag{19}$$

Proposition 9 implies $\tilde{Q}_n$ is a projector on the subspace span $\{\bar{r}_k\}_{k=1}^n \subset \mathbf{E}$. We estimate now the norm $\tilde{Q}_n$.

We recall $X \prec Y$ denotes that

$$\int_0^t X^*(s)ds \leq \int_0^t Y^*(s)ds$$

for each $t \in (0,1)$.

Proposition 10. *Let $\{X_k\}_{k=1}^\infty$ be independent symmetric r.v.s such that $E |X_k| = a > 0$ for all $k \in \mathbf{N}$ and $\{r_k\}_{k=1}^\infty$ be the Rademacher system. Then*

$$\sum_{k=1}^n a_k r_k \prec \sum_{k=1}^n a_k X_k$$

for all $a_k \in \mathbf{R}$ and $n \in \mathbf{N}$.

Proof: Suppose first that $\mu \{X_k = 0\} = 0$ for all k. We put $h_k^+ = \{X_k > 0\}$ and $h_k^- = \{X_k < 0\}$. Let β_k be the σ-algebra generated by $\{h_k^+, h_k^-\}$ and let

$$\tilde{r}_k = a^{-1} E^{\beta_k} X_k .$$

Then the r.v.s $\{\tilde{r}_k\}_{k=1}^n$ are independent and $\tilde{r}_k \overset{d}{=} r_k$.

Let $\beta^{(n)}$ be the σ-algebra generated by $\bigcup_{k=1}^n \beta_k$. By virtue of independence (see [35])

$$E^{\beta^{(n)}} X_k = E^{\beta_k} X_k \quad (1 \leq k \leq n).$$

Therefore

$$\sum_{k=1}^{n} a_k \bar{r}_k = a^{-1} \sum_{k=1}^{n} a_k E^{\beta_k} X_k = a^{-1} E^{\beta^{(n)}} \sum_{k=1}^{n} a_k X_k.$$

The operator $E^{\beta^{(n)}}$ acts in $L_1(0,1)$ and $L_\infty(0,1)$ with the norm 1. Hence $E^{\beta^{(n)}} X \prec X$ (see [29]), which implies the desired relation.

Consider the general case. Each symmetric r.v. X_k may be uniformly approximated by symmetric r.v.s $X_{k,n}$ such that $\mu\{X_{k,n} = 0\} = 0$. We may choose $\{X_{k,n}\}_{k=1}^{\infty}$ to be independent for each integer n. From here and the above, Proposition 10 follows. □

Proposition 11. *Let* **E** *be a separable r.i. space. Then*

$$\left\| \tilde{Q}_n \right\|_{\mathbf{E} \to \mathbf{E}} \leq \frac{\max_{1 \leq k \leq n} \left\| \bar{X}_k \right\|_\infty}{\min_{1 \leq k \leq n} \left\| \bar{X}_k \right\|_1} \left\| Q_n \right\|_{\mathbf{E} \to \mathbf{E}}.$$

Proof: The collections $\{r_k\}_{k=1}^{n}$ and $\{\bar{r}_k\}_{k=1}^{n}$ are equidistributed. From Propositions 8, 10 and 1.3

$$\left\| \sum_{k=1}^{n} a_k \bar{r}_k \right\|_{\mathbf{E}} \leq \left\| \sum_{k=1}^{n} \frac{a_k}{\left\| \bar{X}_k \right\|_1} \bar{X}_k \right\|_{\mathbf{E}}$$

for every $a_k \in \mathbf{R}$. This and (19) give us

$$\left\| \tilde{Q}_n X \right\|_{\mathbf{E}} \leq \left\| \sum_{k=1}^{n} \frac{\langle X, \bar{Y}_k \rangle}{(E\,|\bar{Y}_k|) \left\| \bar{X}_k \right\|_1} \bar{X}_k \right\|_{\mathbf{E}}.$$

Applying Propositions 1.14 and 1.15, we obtain

$$\left\| \tilde{Q}_n X \right\|_{\mathbf{E}} \leq \frac{\left\| \sum_{k=1}^{n} \langle X, \bar{Y}_k \rangle \bar{X}_k \right\|_{\mathbf{E}}}{\left(\min_{1 \leq k \leq n} \left\| \bar{X}_k \right\|_1 \right) \left(\min_{1 \leq k \leq n} E\,|\bar{Y}_k| \right)}.$$

We have $1 = \langle \bar{X}_k, \bar{Y}_k \rangle \leq E\,|\bar{Y}_k| \left\| \bar{X}_k \right\|_\infty$. Hence,

$$\left(\min_{1 \leq k \leq n} E\,|\bar{Y}_k| \right)^{-1} \leq \max_{1 \leq k \leq n} \left\| \bar{X}_k \right\|_\infty.$$

The last relations and (15) imply the needed estimate. □

The next statement was proved in [33] and [49].

Lemma 3. *Let **E** be a separable r.i. space. Suppose there are uniformly bounded projectors $P_n : \mathbf{E} \to \operatorname{span}\{r_k\}_{k=1}^n$ ($n \in \mathbf{N}$). Then each of the spaces **E'** and **E''** contains a r.v. with the normal distribution.*

First we establish two auxiliary statements. Let

$$t = \sum_{k=1}^{\infty} a_k 2^{-k} \quad , \quad s = \sum_{k=1}^{\infty} b_k 2^{-k} \quad (a_k, b_k = 0; 1)$$

be the dyadic decomposition of t, $s \in (0,1)$. Let's put

$$t \dotplus s = \sum_{k=1}^{\infty} 2^{-k} ((a_k + b_k) \mod 2)$$

and consider the transformation $\phi_t(s) = t \dotplus s$, where s, $t \in [0,1]$. It is easy to verify that ϕ_t is an inverse measure-preserving transformation on $[0,1]$.

We put $A_k^+ = \{r_k = 1\}$ and $A_k^- = \{r_k = -1\}$.

Proposition 12. *For each $t \in A_k^+$*

$$\phi_t(A_k^+) = A_k^+ \quad , \quad \phi_t(A_k^-) = A_k^- .$$

If $t \in A_k^-$, then

$$\phi_t(A_k^+) = A_k^- \quad , \quad \phi_t(A_k^-) = A_k^+.$$

Proof: Let $t \in [0,1]$ and a_k be the coefficients of the dyadic decomposition of t. It is easy to verify that the conditions $t \in A_k^+$ and $a_k = 0$ (respectively, $t \in A_k^-$ and $a_k = 1$) are equivalent. This and the definition of the operation $\dotplus$ yield the needed equalities. $\square$

We put $(S(t)X)(s) = X(\phi_t(s))$ ($0 \leq t \leq 1$) and have $S(t_1 \dotplus t_2) = S(t_1)S(t_2)$. The equalities

$$S(t)r_k = r_k \ (t \in A_k^+) \quad , \quad S(t)r_k = -r_k \ (t \in A_k^-) \tag{20}$$

follows from Proposition 12.

Proposition 13. *Let Y be a r.v. such that $S(t)Y = Y$ ($t \in A_k^+$) and $S(t)Y = -Y$ ($t \in A_k^-$). Then $Y = cr_k$ for some constant c.*

Proof: Suppose $t \in A_k^+$. We have $Y(t \dotplus v) = Y(v)$ for every $v \in [0,1]$. Proposition 12 implies $\{t \dotplus v : t \in A_k^+\} = \phi_v(A_k^+) = A_k^+$ for fixed $v \in A_k^+$. Hence, Y is a constant on A_k^+ and, similarly, on A_k^-.

Let $t \in A_k^-$. Then $Y(t \dotplus v) = -Y(v)$ ($0 \leq v \leq 1$). According to Proposition 12, $t \dotplus v \in A_k^-$ if $v \in A_k^-$. Hence

$$Y \Big|_{A_k^-} = -Y \Big|_{A_k^+} .$$

From here $Y = cr_k$, where c is a constant. $\square$

Proof of Lemma 3.: We prove that for the Rademacher system and the spaces **E** and **E'** the estimates (4) hold. From here Lemma 3 follows (see the section 3.2).

For all $t \in [0, 1]$ and $n \in \mathbf{N}$ the operator $S(t)$ maps the subspace $\{r_k\}_{k=1}^n$ into itself, which follows from (20). This operator acts isometrically on **E**. We put

$$Q_n X = \int_0^1 \left(S(t) P_n S(t)^{-1} \right) X \, dt \,.$$

Then Q_n is a projector on the subspace span $\{r_k\}_{k=1}^n$ commuting with each operator $S(t)$ $(0 \le t \le 1)$. In addition $\|Q_n\|_{\mathbf{E} \to \mathbf{E}} \le \|P_n\|_{\mathbf{E} \to \mathbf{E}}$.

Since **E** is separable, there are r.v.s $\{Y_{k,n}\}_{k=1}^n \subset \mathbf{E'}$ such that

$$Q_n X = \sum_{k=1}^n \langle X \,, Y_{k,n} \rangle \, r_k$$

(see the proof of Lemma 2). By virtue of the permutability of Q_n and $S(t)$ and the formulae (20), the r.v.s $Y_{k,n}$ satisfy the conditions of Proposition 13. Hence $Y_{k,n} = c_{k,n} r_k$ and $c_{k,n}$ is a constant. As Q_n is a projector on the subspace span $\{r_k\}_{k=1}^n$, then $\langle r_k \,, Y_{k,n} \rangle = 1$ and $c_{k,n} = 1$ $(1 \le k \le n; n \in \mathbf{N})$.

So, $Q_n X = \sum_{k=1}^n \langle X \,, r_k \rangle \, r_k$. From the assumption of the lemma

$$C \equiv \sup_n \|Q_n\|_{\mathbf{E} \to \mathbf{E}} \le \|P_n\|_{\mathbf{E} \to \mathbf{E}} < \infty \,. \tag{21}$$

It is not hard to verify that $\langle Q_n X \,, Y \rangle = \langle X \,, Q_n Y \rangle$. Since $\mathbf{E'} = \mathbf{E}^*$, we have

$$\|Q_n X\|_{\mathbf{E}} = \sup \left\{ \langle Q_n X \,, Y \rangle : Y \in \mathbf{E'}, \|Y\|_{\mathbf{E'}} \le 1 \right\} \,.$$

These relations give us $\|Q_n\|_{\mathbf{E'} \to \mathbf{E'}} = \|Q_n\|_{\mathbf{E} \to \mathbf{E}}$. Using (21), we get

$$\left\| \sum_{k=1}^n a_k r_k \right\|_{\mathbf{E}} = \left\| Q_n \sum_{k=1}^n a_k r_k \right\|_{\mathbf{E}}$$

$$= \sup \left\{ \left\langle \sum_{k=1}^n a_k r_k \,, Q_n Y \right\rangle : Y \in \mathbf{E'}, \|Y\|_{\mathbf{E'}} \le 1 \right\}$$

$$\le \sup \left\{ \left\langle \sum_{k=1}^n a_k r_k \,, Z \right\rangle : Z \in \text{span} \{r_k\}_{k=1}^n \,, \|Z\|_{\mathbf{E'}} \le C \right\} \,.$$

According to Lemma 3.1, there are positive constants a and b such that for every $a_k \,, b_k \in \mathbf{R}$ and $n \in \mathbf{N}$

$$\left\| \sum_{k=1}^n a_k r_k \right\|_{\mathbf{E}} \ge a \left(\sum_{k=1}^n a^2 \right)^{1/2} , \tag{22}$$

$$\left\|\sum_{k=1}^{n} b_k r_k\right\|_{\mathbf{E}} \geq b\left(\sum_{k=1}^{n} b^2\right)^{1/2}. \tag{23}$$

From here and the above

$$\left\|\sum_{k=1}^{n} a_k r_k\right\|_{\mathbf{E}} \leq \sup\left\{\left\langle\sum_{k=1}^{n} a_k r_k, \sum_{k=1}^{n} b_k r_k\right\rangle : \left\|\sum_{k=1}^{n} b_k r_k\right\|_{\mathbf{E}'} \leq C\right\}$$

$$\leq \sup\left\{\sum_{k=1}^{n} a_k b_k : \left(\sum_{k=1}^{n} b^2\right)^{1/2} \leq b^{-1} C\right\}$$

$$= b^{-1} C\left(\sum_{k=1}^{n} a^2\right)^{1/2}.$$

Similarly

$$\left\|\sum_{k=1}^{n} b_k r_k\right\|_{\mathbf{E}'} \leq a^{-1} C\left(\sum_{k=1}^{n} b_k^2\right)^{1/2}.$$

Therefore, for the sequence $\{r_k\}_{k=1}^{\infty}$ and the spaces $\mathbf{E}$ and $\mathbf{E}'$ the estimates (4) are fulfilled. Reasoning as in the section 3.2 and using the relation $(\mathbf{E}')'' = \mathbf{E}'$ we get that the spaces $\mathbf{E}'$ and $\mathbf{E}''$ contain a r.v. with the normal distribution. □

8. Proof of Theorem 1. Necessity. We may suppose the r.v.s X_k to be symmetric and the r.i. space $\mathbf{E}$ to be separable. We denote by $P : \mathbf{E} \to$ span $\{X_k\}_{k=1}^{\infty}$ a bounded projector and put $R_n X_k = X_k$ for $1 \leq k \leq n$ and $R_n X_k = 0$ for $k > n$. Propositions 1.14 and 1.15 yield that the operator R_n acts in the subspace span $\{X_k\}_{k=1}^{\infty} \subset \mathbf{E}$ with the norm 1. Hence, for the operator $P_n = R_n P$ and all $n \in \mathbf{N}$ we have

$$\|P_n\|_{\mathbf{E}\to\mathbf{E}} \leq \|P\|_{\mathbf{E}\to\mathbf{E}}. \tag{24}$$

We reduce the proof to the case of finite-valued random variables.

Proposition 14. *Let* $\{X_k\}_{k=1}^{n}$ *be independent bounded symmetric r.v.s and* $P : \mathbf{E} \to$ span $\{X_k\}_{k=1}^{n}$ *be a projector. Then for every* $\epsilon > 0$ *there exists a collection of independent symmetric finite-valued r.v.s* $\{X_{k,\epsilon}\}_{k=1}^{n}$ *with the following properties:*
   1)$\|X_k - X_{k,\epsilon}\|_{\mathbf{E}} \leq \epsilon$ $(1 \leq k \leq n)$;
   2) *there is a projector* $P^{(\epsilon)} : \mathbf{E} \to$ span $\{X_{k,\epsilon}\}_{k=1}^{n}$ *such that*

$$(1-\epsilon)\|P\|_{\mathbf{E}\to\mathbf{E}} \leq \|P^\epsilon\|_{\mathbf{E}\to\mathbf{E}} \leq (1+\epsilon)\|P\|_{\mathbf{E}\to\mathbf{E}}.$$

Proof: Since the r.v.s X_k are bounded, we may construct for every integer m a collection of independent finite-valued symmetric r.v.s $\{X_{k,m}\}_{k=1}^n$ such that

$$\|X_k - X_{k,m}\|_\infty \leq m^{-1} \quad (1 \leq k \leq n).$$

From here and Proposition 1.1, $X_{k,m} \to X_k$ and $PX_{k,m} \to PX_k$ in **E** as $m \to \infty$.

We define the operator $J_m : \text{span}\{X_{k,m}\}_{k=1}^n \to \text{span}\{X_k\}_{k=1}^n$ by the equality $J_m X_{k,m} = PX_{k,m}$. For large enough m the operator J_m is invertible and $\|J_m\| \to 1$ and $\|J_m^{-1}\| \to 1$ as $m \to \infty$. We put $P_m = J_m^{-1}P$. Then P_m is a projector on $\text{span}\{X_{k,m}\}_{k=1}^n$ and

$$\|P\|_{\mathbf{E}\to\mathbf{E}} \|J_m\|^{-1} \leq \|P_m\|_{\mathbf{E}\to\mathbf{E}} \leq \|P\|_{\mathbf{E}\to\mathbf{E}} \|J_m^{-1}\| .$$

Choosing m sufficiently large and putting $X_{k,\epsilon} = X_{k,m}$, we get the needed assertion. $\square$

We continue to prove Theorem 1. According to Proposition 14 there exist collections of finite-valued symmetric independent r.v.s $\{X_{k,n}\}_{k=1}^n$ $(n \in \mathbf{N})$ with the following properties:

1) $\|X_{k,n}\|_\infty \leq 2\|X_k\|_\infty$ $(1 \leq k \leq n)$;
2) $\|X_{k,n}\|_1 \geq \frac{1}{2}\|X_k\|_1$ $(1 \leq k \leq n)$;
3) there is a projector $\hat{P}_n : \mathbf{E} \to \text{span}\{X_{k,n}\}_{k=1}^n$ such that

$$\left\|\hat{P}\right\|_{\mathbf{E}\to\mathbf{E}} \leq 2\|P\|_{\mathbf{E}\to\mathbf{E}} .$$

Let the r.v.s $\{\bar{X}_{k,n}\}_{k=1}^n$ be constructed by the formula (12) and the r.v.s $\{X_{k,n}\}_{k=1}^n$. Consider the r.v.s $\{\bar{r}_k\}_{k=1}^n$ determined by the formula (18). According to Lemma 2 and Proposition 10 there is a projector $\tilde{Q}_n : \mathbf{E} \to \text{span}\{\bar{r}_k\}_{k=1}^n$ such that

$$\left\|\tilde{Q}_n\right\|_{\mathbf{E}\to\mathbf{E}} \leq \frac{\max_{1\leq k\leq n}\|\bar{X}_{k,n}\|_\infty}{\min_{1\leq k\leq n}\|\bar{X}_{k,n}\|_1} \left\|\hat{P}\right\|_{\mathbf{E}\to\mathbf{E}} .$$

Since $X_{k,n} \overset{d}{=} X_{k,n}$, then the assumptions of Theorem 1, the estimate (24) and the properties 1)—3) imply

$$\left\|\tilde{Q}_n\right\|_{\mathbf{E}\to\mathbf{E}} \leq 8\frac{\max_{1\leq k<\infty}\|X_k\|_\infty}{\min_{1\leq k<\infty}\|X_k\|_1} \|P\|_{\mathbf{E}\to\mathbf{E}} .$$

According to (18) and Proposition 8, the collections $\{\bar{r}_k\}_{k=1}^n$ and $\{r_k\}_{k=1}^n$ are equidistributed. Proposition 5 yields that there is a sequence of uniformly bounded projectors from **E** on the subspaces $\text{span}\{r_k\}_{k=1}^n$ $(n \in \mathbf{N})$. Applying Lemma 3 we obtain Theorem 1. $\square$

2. Subspaces generated by independent equidistributed random variables

1. Result. In this section we consider i.i.d.r.v.s $\{X_k\}_{k=1}^{\infty} \subset \mathbf{E}$ and study the conditions of complementability of the corresponding subspace.

Theorem 2. *Let* $\mathbf{E}$ *be a separable r.i. space and* $\{X_k\}_{k=1}^{\infty} \subset \mathbf{E}$ *be i.i.d.r.v.s. The subspace* span $\{X_k\}_{k=1}^{\infty} \subset \mathbf{E}$ *is complemented if and only if the following hold:*

1) each of the spaces $\mathbf{E}'$ *and* $\mathbf{E}''$ *contains a r.v. with the normal distribution;*

2) the considered subspace is isomorphic to l_2 .

We examine more explicitly the condition 2). We may assume that $EX_1 = 0$. According to Proposition 1.14 the sequence $\{X_k\}_{k=1}^{\infty}$ is the unconditional basis in the subspace span $\{X_k\}_{k=1}^{\infty} \subset \mathbf{E}$. From Theorem 1.4 $\{X_k\}_{k=1}^{\infty}$ is equivalent to the orthogonal basis in Hilbert space. Hence, if $EX_1 = 0$, then 2) is equivalent to the estimate which has been considered in the section 3.2.

2. Proof of Theorem 2. Sufficiency. Without loss of generality $EX_1 = 0$. We denote $h_k^+ = \{X_k \geq 0\}$ and $h_k^- = \{X_k < 0\}$. Each of these sets has a positive measure. We put

$$Y_k = I_{h_k^+} - aI_{h_k^-} , \tag{25}$$

where $a > 0$ is chosen under the condition $EY_k = 0$. Then the r.v.s $\{Y_k\}_{k=1}^{\infty}$ are independent and equidistributed,

$$E(X_j Y_k) = 0 \ (j \neq k) \quad , \quad E(X_k Y_k) = b > 0 , \tag{26}$$

where b is independent of k.

The sequence $\{Y_k\}_{k=1}^{\infty}$ and the space $\mathbf{E}'$ satisfy the conditions of Theorem 1. Hence the estimates (4) holds, i.e. there is a positive constant B such that for every $b_k \in \mathbf{E}$ and $n \in \mathbf{N}$

$$B^{-1} \left(\sum_{k=1}^{n} b_k^2 \right)^{1/2} \leq \left\| \sum_{k=1}^{n} b_k Y_k \right\|_{\mathbf{E}'} \leq B \left(\sum_{k=1}^{n} b_k^2 \right)^{1/2} . \tag{27}$$

We put

$$QX = \frac{1}{b} \sum_{k=1}^{\infty} \langle X , Y_k \rangle X_k .$$

Reasoning as in the proof of Theorem 1 and using (26) and (27), we get that Q is a bounded projector on the considered subspace. $\square$

3. The construction of the special projectors. We turn to the proof of necessity. According to Lemma 1, we may assume the r.v.s X_k to be symmetric.

The scheme of the proof is the following. First we construct bounded symmetric i.i.d.r.v.s $\{V_k\}_{k=1}^{\infty}$ such that there exists a sequence of uniformly bounded projectors from $\mathbf{E}$ on the subspaces $\mathrm{span}\{V_k\}_{k=1}^{n}$ ($n \in \mathbf{N}$). As above, this yields the condition 1). Applying 1), we obtain the condition 2).

Consider first a collection $\{X_k\}_{k=1}^{n}$ of symmetric finite-valued i.i.d.r.v.s. Denote the group of all linear operators on $\mathbf{R}^n$, which rearrange the coordinates of vectors and change their signs, by W^n. More explicitly, let S^n be the group of all permutations of the set $\{1,\ldots,n\}$ (the symmetric group). For each $g \in W^n$ there is a permutation $\sigma_g \in S^n$ and numbers $\epsilon_k(g) = \pm 1$ such that

$$g(\mathbf{v}) = \left(\epsilon_1(g)v_{\sigma_g(1)},\ldots,\epsilon_n(g)v_{\sigma_g(n)}\right), \tag{28}$$

where $\mathbf{v} = (v_1,\ldots,v_n)$.

Let U_n be defined by (10). Since the r.v.s $\{X_k\}_{k=1}^{n}$ are equidistributed, the sets Θ_{m_k} are identical. From here $g(U_n) = U_n$ for every $g \in W^n$.

As in the section 1, we construct the collection of segments $\Delta(\mathbf{v})$, $\mathbf{v} \in U_n$, with the properties described in Proposition 2. Using the formulae (12) and (13) we define the r.v.s $\{\bar{X}_k\}_{k=1}^{n}$ and the invertible measure-preserving transformations ϕ_g ($g \in W^n$). As above, $T(g)$ are the corresponding operators.

Proposition 15. *The following relations are true:*

$$T(g)\bar{X}_k = \epsilon_k(g)\bar{X}_{\sigma_g(k)} \quad (1 \leq k \leq n,\, g \in W^n).$$

The proof is similar to the proof of Proposition 6 . One has to take into account that the r.v.s $\{\bar{X}_k\}_{k=1}^{n}$ are equidistributed, which yields that

$$a^{(k)}(v_{\sigma_g(k)}) = a^{(\sigma_g(k))}(v_{\sigma_g(k)})$$

for each $\mathbf{v} = (v_1,\ldots,v_n) \in U_n$. Hence,

$$\bar{X}_k\Big|_{\Delta(g(\mathbf{v}))} = \epsilon_k(g)\bar{X}_{\sigma_g(k)}\Big|_{\Delta(\mathbf{v})}.$$

Let the r.v.s $\{\bar{Y}_k\}_{k=1}^{n} \subset \mathbf{E}'$ define the projector from $\mathbf{E}$ on $\mathrm{span}\{\bar{X}_k\}_{k=1}^{n}$ by the formula

$$\bar{Q}X = \sum_{k=1}^{n}\langle X, \bar{Y}_k\rangle\bar{X}_k \tag{29}$$

and suppose that $\bar{Q}$ commutes with all operators $T(g)$, $g \in W^n$. The equality (16) holds for W^n. Applying Proposition 15, we easily verify that

$$T(g)\bar{Y}_k = \epsilon_k(g)\bar{Y}_{\sigma_g(k)} \quad (1 \leq k \leq n,\, g \in W^n). \tag{30}$$

The next step is to construct a projector on the subspace $\mathrm{span}\{X_k\}_{k=1}^{n}$ with some special properties.

Proposition 16. *Let* $\{X_k\}_{k=1}^n$ *be finite-valued symmetric i.i.d.r.v.s,* $\mathbf{E}$ *be a separable r.i. space and* $P_n : \mathbf{E} \to \operatorname{span} \{X_k\}_{k=1}^n$ *be a projector. Then there are r.v.s* $\{Y_k\}_{k=1}^n \subset \mathbf{E}'$ *with the following properties:*

1) $\langle X_j , Y_k \rangle = \delta_{j,k}$;

2) $E\left(Y_k I_{\{a \leq X_j \leq b\}}\right) = 0$ *for all* $a < b$ *and* $j \neq k$;

3) *the variable* $E\left(Y_k I_{\{a \leq X_k \leq b\}}\right)$ *is independent of* k ;

4) *for the projector*

$$Q_n X = \sum_{k=1}^n \langle X , Y_k \rangle X_k$$

the estimate $\|Q_n\|_{\mathbf{E} \to \mathbf{E}} \leq \|P_n\|_{\mathbf{E} \to \mathbf{E}}$ *holds* .

Proof: According to Proposition 5 there is a projector

$$\bar{P}_n : \mathbf{E} \to \operatorname{span} \left\{\bar{X}_k\right\}_{k=1}^n$$

such that $\left\|\bar{P}_n\right\|_{\mathbf{E} \to \mathbf{E}} \leq \|P_n\|_{\mathbf{E} \to \mathbf{E}}$, where the r.v.s $\bar{X}_k$ are determined by the formula (12). We put

$$\bar{Q}_n X = \frac{1}{2^n n!} \sum_{g \in W^n} \left(T(g)\bar{P}_n T\left(g^{-1}\right)\right) X.$$

It follows from Proposition 15 that every operator $T(g)$ $(g \in W^n)$ maps the span $\left\{\bar{X}_k\right\}_{k=1}^n$ into itself. Since the group W^n has $2^n n!$ elements, $\bar{Q}_n$ is a projector on this subspace and

$$\left\|\bar{Q}_n\right\|_{\mathbf{E} \to \mathbf{E}} \leq \left\|\bar{P}_n\right\|_{\mathbf{E} \to \mathbf{E}} \leq \|P_n\|_{\mathbf{E} \to \mathbf{E}}$$

In addition, $\bar{Q}_n$ commutes with every $T(g)$ $(g \in W^n)$. As $\mathbf{E}$ is separable, $\bar{Q}_n$ is represented in the form (29) and 1) is fulfilled.

Let $k \neq j$. Using (28) we may choose $g \in W^n$ so that $\epsilon_k(g) = 1$, $\epsilon_j(g) = -1$ and $\sigma_g(m) = m$ $(1 \leq m \leq n)$. From (3) and Proposition 15

$$T(g)\bar{Y}_k = -\bar{Y}_k \quad , \quad T(g)\bar{X}_j = \bar{X}_j .$$

So for all $a < b$

$$T(g)I_{\{a \leq \bar{X}_j \leq b\}} = I_{\{a \leq T(g)\bar{X}_j \leq b\}} = I_{\{a \leq \bar{X}_j \leq b\}} .$$

Since $g = g^{-1}$, then from here and (16)

$$E\left(\bar{Y}_k I_{\{a \leq \bar{X}_j \leq b\}}\right) = \left\langle T(g)I_{\{a \leq \bar{X}_j \leq b\}} , \bar{Y}_k \right\rangle$$

$$= \left\langle I_{\{a \leq \bar{X}_j \leq b\}} , T(g)\bar{Y}_k \right\rangle = -E\left(\bar{Y}_k I_{\{a \leq \bar{X}_j \leq b\}}\right) .$$

This implies 2).

Now we choose $g \in W^n$ such that $\epsilon_m(g) = 1$ $(1 \leq m \leq n)$ and $\sigma_g(k) = j$ for fixed $j \neq k$. From (30) and Proposition 15, $T(g)\bar{X}_k = \bar{X}_j$ and $T(g)\bar{Y}_k = \bar{Y}_j$. Since ϕ_g is measure-preserving, the distributions of the pairs $(\bar{X}_k, \bar{Y}_k)$ and $(\bar{X}_j, \bar{Y}_j)$ are identical and 3) follows.

According to Proposition 4, $\bar{X}_k = X_k(\psi_n(t))$ $(1 \leq k \leq n)$, where $\psi_n : [0,1] \longrightarrow [0,1]$ is a measure-preserving transformation. Putting $Y_k(t) = \bar{Y}_k(\psi_n^{-1}(t))$ we get the r.v.s with the needed properties. $\square$

Now we eliminate the assumption that X_k is finite-valued.

Proposition 17. *Let r.i. the space* **E** *be separable and* $\{X_k\}_{k=1}^n \subset \mathbf{E}$ *be symmetric i.i.d.r.v.s. Let* P_n *be a projector on the subspace* span $\{X_k\}_{k=1}^n$. *Then there are r.v.s* $\{Y_k\}_{k=1}^n \subset \mathbf{E}'$ *such that the relations 1)—4) of Proposition 16 are valid.*

Proof: Since **E** is separable, there exist finite-valued r.v.s $X_{k,m}$ $(1 \leq k \leq n ; m \in \mathbf{N})$ with the following properties:

(i) for every m the r.v.s $\{X_{k,m}\}_{k=1}^n$ are symmetric, independent and equidistributed;

(ii) $X_{k,m} \to X_k$ in **E** as $m \to \infty$ $(1 \leq k \leq n)$.

As in the proof of Proposition 14, we construct a collection of projectors $P_{n,m} : \mathbf{E} \to \text{span}\{X_{k,m}\}_{k=1}^n$ such that

$$\lim_{m \to \infty} \|P_{n,m}\|_{\mathbf{E} \to \mathbf{E}} = \|P_n\|_{\mathbf{E} \to \mathbf{E}} . \tag{31}$$

Proposition 16 yields that for each m there are r.v.s $\{Y_{k,m}\}_{k=1}^n \subset \mathbf{E}'$ having the properties 1)—4). Specifically, for the projectors

$$Q_{n,m} X = \sum_{k=1}^n \langle X, Y_{k,m} \rangle X_{k,m}$$

the estimate

$$\|Q_{m,n}\|_{\mathbf{E} \to \mathbf{E}} \leq \|P_{m,n}\|_{\mathbf{E} \to \mathbf{E}}$$

holds. From Propositions 1.14 and 1.15

$$|\langle X, Y_{k,m} \rangle| \|X_{k,m}\|_{\mathbf{E}} \leq \|Q_{n,m} X\|_{\mathbf{E}} \quad (1 \leq k \leq n).$$

As $\|X_{k,m}\|_{\mathbf{E}} \to \|X_k\|_{\mathbf{E}}$ $(m \to \infty)$, then it follows that

$$\sup_m \|Y_{k,m}\|_{\mathbf{E}'} < \infty \quad (1 \leq k \leq n).$$

Using the well known results about the weak compactness [16] and the relation $\mathbf{E}' = \mathbf{E}^*$, we obtain that there are r.v.s $\{Y_k\}_{k=1}^n \subset \mathbf{E}'$ and integers $m(j) \nearrow \infty$ such that

$$\lim_{j \to \infty} \langle X, Y_{k,m(j)} \rangle = \langle X, Y_k \rangle \quad (1 \leq k \leq n) \tag{33}$$

for every $X \in \mathbf{E}$.

We show that the r.v.s $\{Y_k\}_{k=1}^n$ have the desired properties. It follows from (33) and the condition *(ii)* that

$$\langle X_i \,, Y_k \rangle = \lim_{j \to \infty} \langle X_{i,m(j)} \,, Y_{k,m(j)} \rangle = \delta_{j,k}.$$

To prove the properties 2) and 3) we show that

$$\lim_{j \to \infty} E\left(Y_{k,m(j)} I_{\{a \le X_{i,m(j)} < b\}}\right) = E\left(Y_k I_{\{a \le X_i < b\}}\right) \qquad (34)$$

for every $1 \le i$, $k \le n$ and $a < b$.

Proposition 1.1 implies $X_{k,m} \to X_k$ in the space $L_1(0,1)$. Therefore, passing to a subsequence, we may suppose that $X_{k,m} \to X_k$ almost surely [35] . From here for every $a < b$

$$I_{\{a \le X_{i,m(j)} < b\}} \to I_{\{a \le X_i < b\}} \quad (1 \le k \le n) \qquad (35)$$

almost surely as $j \to \infty$. According to (33),

$$\lim_{j \to \infty} E\left(Y_{k,m(j)} I_h\right) = E\left(Y_k I_h\right)$$

for each measurable $h \subset [0,1]$. Using the Vitali—Hahn—Saks theorem (see [16]) we conclude that $E\left(Y_{k,m(j)} I_h\right) \to 0$ uniformly with respect to j and k as $\mu h \to 0$. From here and (35) the relation (34) follows. Using Proposition 16 we get the relations 2) and 3).

We put $Q_n X = \sum_{k=1}^n \langle X \,, Y_k \rangle X_k$. From (33) $Q_{n,m(j)} X \to Q_n X$ in $\mathbf{E}$. Taking into account (31) and (32) we obtain $\|Q_n\|_{\mathbf{E} \to \mathbf{E}} \le \|P_n\|_{\mathbf{E} \to \mathbf{E}}$.□

Lastly we consider an infinite sequence of i.i.d. random variables.

Lemma 4. *Suppose* $\mathbf{E}$ *is a separable r.i. space,* $\{X_k\}_{k=1}^\infty \subset \mathbf{E}$ *are symmetric i.i.d.r.v.s and there exists a family of uniformly bounded projectors* $P_n : \mathbf{E} \to$ span $\{X_k\}_{k=1}^n$ $(n \in \mathbf{N})$. *Then there exist r.v.s* $\{Y_k\}_{k=1}^\infty \subset \mathbf{E}'$ *such that the relations 1)—4) of Proposition 16 are valid for each* n.

Proof: Fix n and let $\{Y_{k,n}\}_{k=1}^n$ be the r.v.s constructed in Proposition 17 for the collection $\{X_k\}_{k=1}^n$. Then for the projectors $\tilde{Q}_n X = \sum_{k=1}^n \langle X \,, Y_{k,n} \rangle X_k$ the relation

$$\sup_n \left\| \tilde{Q}_n \right\|_{\mathbf{E} \to \mathbf{E}} \le \|P_n\|_{\mathbf{E} \to \mathbf{E}} = C < \infty$$

holds. Propositions 1.14 and 1.15 imply for every $X \in \mathbf{E}$

$$|\langle X \,, Y_{k,n} \rangle| \, \|X_1\|_{\mathbf{E}} = |\langle X \,, Y_{k,n} \rangle| \, \|X_k\|_{\mathbf{E}} \le \left\| \tilde{Q}_n X \right\|_{\mathbf{E}} \le C \, \|X\|_{\mathbf{E}} \,,$$

which gives us

$$\|Y_{k,n}\|_{\mathbf{E}'} \leq \frac{C}{\|X_1\|_{\mathbf{E}}} \quad (1 \leq k \leq n, \, n \in \mathbf{N}).$$

Reasoning as in the proof of Proposition 17 and using the diagonal process, we may find integers $n(j) \nearrow \infty$ and r.v.s $\{Y_k\}_{k=1}^{\infty} \subset \mathbf{E}'$ such that for every $X \in \mathbf{E}$ and $k \in \mathbf{N}$

$$\lim_{j \to \infty} \langle X, Y_{k,n(j)} \rangle = \langle X, Y_k \rangle \, .$$

The r.v.s $\{Y_k\}_{k=1}^{\infty}$ have the desired properties. $\square$

4. Reduction to Theorem 1. Now we construct bounded symmetric i.i.d.r.v.s $\{V_k\}_{k=1}^{\infty}$ for which there is a sequence of uniformly bounded projectors from $\mathbf{E}$ to the subspaces span $\{V_k\}_{k=1}^{n}$. We put

$$V_{k,c} = X_k I_{\{|X_k| \leq c\}} \, , \tag{36}$$

where $c > 0$ is a constant.

Lemma 5. *Let the conditions of Lemma 4 hold. Then for large enough c there is a collection of uniformly bounded projectors on the subspaces* span $\{V_{k,c}\}_{k=1}^{n}$.

Proof: Let $\{Y_k\}_{k=1}^{\infty} \subset \mathbf{E}'$ be the r.v.s constructed in Lemma 4. The property 2) implies $\langle V_{k,c}, Y_j \rangle = 0$ for $j \neq k$. According to the property 3) the quantity

$$h(c) = \langle V_{k,c}, Y_k \rangle \tag{37}$$

is independent of k. We have $h(c) \to \langle X_k, Y_k \rangle = 1$ as $c \to \infty$. Hence, $h(c) > 1/2$ for large enough c. We put

$$Q_{n,c}X = \frac{1}{h(c)} \sum_{k=1}^{n} \langle X, Y_k \rangle V_{k,c} \tag{38}$$

and show the projectors $Q_{n,c}$ are uniformly bounded.

Let's denote

$$X_{k,c} = X_k I_{\{|X_k| \leq c\}} - X_k I_{\{|X_k| > c\}} \, .$$

By virtue of symmetry $X_{k,c} \stackrel{d}{=} X_k$. Hence, the sums

$$Q_n X = \sum_{k=1}^{n} \langle X, Y_k \rangle X_k \quad , \quad R_{n,c}X = \sum_{k=1}^{n} \langle X, Y_k \rangle X_{k,c}$$

are equidistributed and $\|Q_n X\|_{\mathbf{E}} = \|R_{n,c} X\|_{\mathbf{E}}$ for each $X \in \mathbf{E}$. From (36) and (38)

$$Q_{n,c} X = \frac{1}{2h(c)} \left(Q_n X + R_{n,c} X \right),$$

which yields

$$\|Q_{n,c} X\|_{\mathbf{E}} \leq \frac{\|Q_n X\|_{\mathbf{E}}}{h(c)} \leq 2 \|Q_n\|_{\mathbf{E} \to \mathbf{E}} \|X\|_{\mathbf{E}}.$$

Lemma 5 follows from here and Lemma 4. $\square$

5. Proof of Theorem 2. Necessity. We may assume as mentioned above, the r.v.s X_k to be symmetric. It follows from the complementability of span $\{X_k\}_{k=1}^{\infty}$ that there exists a sequence of uniformly bounded projectors on the subspaces span $\{X_k\}_{k=1}^{n}$. Using Lemma 5 and applying to the r.v.s $\{V_{k,c}\}_{k=1}^{\infty}$ the arguments of subsection 8 of the section 1, we get the assertion 1) of Theorem 2.

Now we show that span $\{X_k\}_{k=1}^{\infty} \subset \mathbf{E}$ is isomorphic to the space l_2. According to Lemma 3.1

$$\left\| \sum_{k=1}^{n} a_k X_k \right\|_{\mathbf{E}} \geq a \left(\sum_{k=1}^{n} a^2 \right)^{1/2},$$

where $a > 0$ is independent of n and a_k. So, we have to prove the corresponding upper estimate.

For the r.v.s $\{V_{k,c}\}_{k=1}^{\infty}$ the inequality (4) holds, i.e. there is a positive constant $D(c)$ such that

$$D(c)^{-1} \left(\sum_{k=1}^{n} a^2 \right)^{1/2} \leq \left\| \sum_{k=1}^{n} a_k V_{k,c} \right\|_{\mathbf{E}} \leq D(c) \left(\sum_{k=1}^{n} a^2 \right)^{1/2}$$

for all integers n and $a_k \in \mathbf{R}$. Let $\{Y_k\}_{k=1}^{\infty} \subset \mathbf{E}'$ be the r.v.s constructed in Lemma 4. We choose $c > 0$ under the condition $h(c) > 1/2$, where $h(c)$ is determined by the formula (37). The last bound gives us

$$\left\| \sum_{k=1}^{n} b_k Y_k \right\|_{\mathbf{E}'} \geq \sup \left\{ \left\langle \sum_{k=1}^{n} b_k Y_k, X \right\rangle : X \in \text{span} \{V_{k,c}\}_{k=1}^{n}, \|X\|_{\mathbf{E}} \leq 1 \right\}$$

$$\geq \sup \left\{ h(c) \sum_{k=1}^{n} a_k b_k : \left(\sum_{k=1}^{n} a^2 \right)^{1/2} \leq \frac{1}{D(c)} \right\}$$

$$\geq \frac{1}{2D(c)} \left(\sum_{k=1}^{n} b^2 \right)^{1/2}. \tag{39}$$

Now we can prove the desired upper estimate. Let Q_n be the projectors constructed in Lemma 4. We have

$$C \equiv \sup_n \|Q_n\|_{\mathbf{E}\to\mathbf{E}} < \infty.$$

Since $\mathbf{E}$ is separable, then

$$\|X\|_{\mathbf{E}} = \sup\left\{\langle X, Y\rangle : Y \in \mathbf{E}', \|Y\|_{\mathbf{E}'} \le 1\right\}.$$

It is not difficult to verify that the conjugate operator Q_n^* is the projector from $\mathbf{E}'$ on the subspace span $\{Y_k\}_{k=1}^n$ and the equality

$$\|Q_n^*\|_{\mathbf{E}'\to\mathbf{E}'} = \|Q_n\|_{\mathbf{E}\to\mathbf{E}}$$

holds. Using the relation (39), we get

$$
\left\|\sum_{k=1}^n a_k X_k\right\|_{\mathbf{E}} = \left\|Q_n\sum_{k=1}^n a_k X_k\right\|_{\mathbf{E}}
$$

$$
= \sup\left\{\left\langle \sum_{k=1}^n a_k X_k, Q_n^* Y\right\rangle : Y \in \mathbf{E}', \|Y\|_{\mathbf{E}'} \le 1\right\}
$$

$$
\le \sup\left\{\sum_{k=1}^n a_k b_k : \left\|\sum_{k=1}^n b_k Y_k\right\|_{\mathbf{E}'} \le \|Q_n^*\|_{\mathbf{E}'\to\mathbf{E}'}\right\}
$$

$$
\le \sup\left\{\sum_{k=1}^n a_k b_k : \left(\sum_{k=1}^n b^2\right)^{1/2} \le 2D(c)\|Q_n^*\|_{\mathbf{E}'\to\mathbf{E}'}\right\}
$$

$$
= 2CD(c)\left(\sum_{k=1}^n a^2\right)^{1/2},
$$

which completes the proof. □

REFERENCES

1. A. de Acosta, *Exponential moments of vector valueded random series and trianguar series*, Ann. Prob. **8** (1980), 381–389.
2. B. von Bahr and K.-G. Esseen, *Inequalities for the rth absolute moment of a sum of independent random variables, $1 \leq r \leq 2$*, Ann. Math. Statist. **36** (1965), 299–303.
3. S.N. Bernstein, *Some remarks about Liapunov's limit theorem*, Dokl. AN SSSR (Russian) **24** (1939), 3–7.
4. K.G. Binmore and H.H. Stratton, *A note on characteristic functions*, Ann. Math. Statist. **40** (1969), 303–307.
5. R.P. Boas, *Lipschitz behavior and integrability of characteristic functions*, Ann. Math. Statist. **38** (1967), 32–36.
6. M.Sh. Braverman, *Complementability of subspaces generated by independent functions in a symmetric space*, Funct. Anal. Prilozen. (Russian) **(16)** **2** (1982), 66–67.
7. *Random variables with infinitely divisible distributions and symmetric spaces*, Sib. Math. J. (Russian) **(26)** **2** (1985), 36–50.
8. *Symmetric spaces and sequences of independent random varibles*, Funct. Anal. Prilozen. (Russian) **(19)** **4** (1985), 78–79.
9. *On a condition for absolute moments of sums of independent random variables*, Ukranian Math. J. (Russian) **(41)** **11** (1989), 1450–1455.
10. *On symmetric spaces and sequences of indepenedent random variables*, Theor. Veroyatn. Prim. (Russian) **(35)** **4** (1989), 561–565.
11. *Rosenthal's inequality and a characterization of the spaces L_p*, Sib. Math. J. (Russian) **(31)** **3** (1991), 31–38.
12. *Exponential Orlicz spaces and independent random variables*, Prob. and Math. Statist. **(12)** **2** (1991), 245–251.
13. M.Sh. Braverman and I. Ya. Novikov, *Subspaces of symmetric spaces generated by independent random variables*, Sib. Math. J. (Russian) **(25)** **3** (1984), 31–39.
14. N.L. Carothers and S.J. Dilworth, *Inequalities for sums of independent random variables*, Proc. Amer. Math. Soc. **(104)** (1988), 221–226.
15. *Equidistributed random variables in $L_{p,q}$*, J. Funct. Anal. **84** (1989), 146–159.
16. N. Danford and J.T. Schwartz, "Linear Operators. Part I: General Theory," Interscience Publishers, New York and London, 1958.
17. L.E. Dor and T. Stabird, *Projections of L_p onto subspaces spanned by independent random variables*, Compositio Math. **39** (1979), 249–274.

18. K.-G. Esseen and S. Janson, *On moment conditions for normed sums of independent random variables and martingale differences*, Stoch. Processes Appl. **9** (1985), 173–185.

19. W. Feller, "An Introduction to Probability Theory and its Applications," 2nd ed., Wiley, 1970.

20. "An Introduction to Probability Theory and its Applications," 2nd ed., Wiley, 1971.

21. I.M. Gelfand, *Remarks on the work of N.K. Bari "Biorthogonal systems and bases in Hilbert spaces"*, Moscow Gos. Univ. Uch. Zap. (Russian) **(148) 4** (1951), 224–225.

22. I.A. Ibragimov and Yu. V. Linnik, "Independent and Stationary Dependent Random Variables," Wolters-Noordhoff, Groningen, 1971.

23. W.B. Johnson, B. Maurey, G. Schechtman and L. Tzafriri, *Symmetric Structures in Banach Spaces*, Mem. Amer. Math. Soc. **217** (1979), 1–298.

24. W.B. Johnson and G. Schechtman, *Sums of independent random variables in rearrangement invariant spaces*, Ann. Prob. **17** (1989), 789–808.

25. M.J. Kadec and A. Pelczynski, *Bases, lacunary sequences and complemented subspaces in the spaces L_p*, Studia Math. **21** (1962), 161–176.

26. J.P. Kahane, "Some Random Series of Functions.," D.C. Heath and Company, Lexington, Massachusetts, 1968.

27. B.S. Kashin and A.A. Saakyan, "Orthogonal series," Nauka (Russian), Moscow, 1984.

28. M.A. Krasnoselskii and Ya.B. Rutickii, "Convex Functions and Orlich Spaces," Fizmatgiz (Russian), Moscow, 1958.

29. S.G. Krein, Yu.I. Petunin and E.M. Semenov, "Interpolation of Liner Operators,," Trans. Math. Monograph **54**, Amer. Math. Soc., Providence, 1982.

30. V.M. Kruglov, *Remark on the theory of infinitely divisible laws*, Teor. Veroyatn. Prim. (Russian) **(15) 2** (1970), 331–336.

31. V.M. Kruglov and S.N. Antonov, *On the asymptotic behavior of infinite divisible distributions on Banach spaces*, Teor. Veroyatn. Prim. (Russian) **(27) 4** (1982), 625–642.

32. J. Lindenstrauss and L. Tzafriri, "Classical Banach Spaces, vol 1. Sequence Spaces," Springer-Verlag, Berlin, 1977.

33. "Classical Banach Spaces, vol 2. Function Spaces," Springer-Verlag, Berlin, 1979.

34. Yu.V. Linnik and N.V. Ostrovskii, "Decompositions of Random Variables and Vectors," Nauka (Russian), Moscow, 1970.

35. M. Loev, "Probability Theory," 2nd rev. ed., Van Nostrand, Princeton, N.J., 1960.

36. E. Lukacz, "Characteristic Functions," 2nd ed., Griffin, New York and London, 1970.

37. M.B. Marcus and G. Pisier, *Stochastic processes with sample path in exponenthial Orlicz spaces*, in "Lect. Notes Math.," Springer-Verlag, Berlin, 1985, pp. 329–358.

38. V.D. Milman, *Geometrical theory of Banach spaces*, Uspechi Math. Nauk (Russian) **25** (1970), 113–174.

39. S.Ya. Novikov, *Cotype and type of function Lorentz spaces*, Math. Za-metki (Russian) **(32)** 2 (1982), 213–221.

40. S.Ya. Novikov, E.M. Semenov and E.V. Tokarev, *A structure of sub-spaces of the space* $\Lambda_p(\phi)$, Dokl. AN SSSR (Russian) **(247)** 3 (1979)), 552–554.

41. A. Pelczynski, *On the isomorphism of the spaces* **m** *and* **M**, Bull. Acad. Pol. Sci., Serie math., astr. and phys. 6 (1956), 695–696.

42. *Projections in certain Banach spaces*, Studia Math. **19** (1960), 209–228.

43. V.V. Petrov, "Sums of Independent Random Variables," Springer-Verlag, Berlin, 1975.

44. G. Pisier, *Probabilistic methods in the geometry of Banach space*, in "Lect. Notes Math.," Springer-Verlag, Berlin, 1986, pp. 167–241.

45. Yu. V. Prokhorov, *Strong stability of sums and infinitely divisible laws*, Teor. Veroyat. Prim. (Russian) **(3)** 2 (1958), 153–165.

46. *Some extremal problem of probability theory*, Teor. Veroyatn. Prim. (Russian) **(4)** 2 (1959), 211–214.

47. R. Pyke and D. Root, *On convergence in r-mean of normalized partial sums*, Ann. Math. Statist. 39 (1968), 379–381.

48. V.A. Rodin and E.M. Semenov, *Rademacher series in symmetric spaces*, Analysis Math. 1 (1975), 202–222.

49. *The complementability of a subspace that is generated by Radema- cher system in symmetric spaces*, Funct. Anal. Prilozen. (Russian) **(13)** 2 (1979), 91–92.

50. H.P. Rosenthal, *On the subspaces of L_p (p > 2) spanned by sequences of independent random variables*, Israel J. Math. 8 (1970), 273–303.

51. A.N. Shyriaev, "Probability (Russian)," Nauka, Moscow, 1980.

52. E.M. Stein and G. Weiss, "Introduction to Fourier Analysis on Euclidean Spaces," Princeton University Press, Princeton, N.J., 1971.

53. Vakhania, N.N., Tarieladze, V.I. and Chobanian, S.A., "Probability Dis-tributions on Banach Spaces," Reidel, Dordrecht, 1987.

54. M. Zippin, *On perfectly homogeneous bases in Banach spaces*, Israel J. Math. 4 (1966), 265–272.

55. V.M. Zolotarev, "One-dimentional Stable Distributions," Nauka (Rus-sian), Moscow, 1983.

Index

Printed in the United States
By Bookmasters